AF553702

Growth and Development of AGRICULTURE

Editor
Ramniwas Sharma

2006
Biotech Books
Delhi - 110 035

ISBN 81-7622-162-7

Published by : **BIOTECH BOOKS**
1123/74, Tri Nagar,
DELHI – 110 035
Phone: 27382765
e-mail: biotechbooks@yahoo.co.in

Showroom : 4762-63/23, Ansari Road, Darya Ganj,
NEW DELHI - 110 002
Phone: 23245578, 23244987

Printed at : Tarun Offset Printers,

Preface

Agronomy today is very different from what it was before about 1950. Intensification of agriculture since the 1960s in developed and developing countries, often referred to as the Green Revolution, was closely tied to progress made in selecting and improving crops and animals for high productivity, as well as to developing additional inputs such as artificial fertilizers and phytosanitary productss.

Agriculture is the process of producing food, feed, fiber and other desired products by the cultivation of certain plants and the raising of domesticated animals. The practice of agriculture is also known as farming, while scientists, inventors and others devoted to improving farming methods and implements are also said to be engaged in agriculture. More people in the world are involved in agriculture as their primary economic activity than in any other, yet it only accounts for four percent of the world's

Agriculture can refer to subsistence agriculture, the production of enough food to meet just the needs of the farmer/agriculturalist and his/her family. It may also refer to industrial agriculture, long prevalent in developed nations and increasingly so elsewhere, which consists of obtaining financial income from the cultivation of land to yield produce, the commercial raising of animals, or both. Agriculture is also short for the study of the practice of agriculture-more formally known as agricultural science.

Modern agriculture depends heavily on engineering and technology and on the biological and physical sciences. Irrigation, drainage, conservation and sanitary engineering, each of which is important in successful farming, are some of the fields requiring the specialized knowledge of agricultural engineers.

Agricultural chemistry deals with other vital farming concerns, such as the application of fertilizer, insecticides, and fungicides, soil makeup, analysis of agricultural products, and nutritional needs of farm animals.

The book has a lucid, expository style. It reviews the various theories over the years to explain why some countries are rich while others are poor. It has been assumed that all human populations have equal mental abilities to adopt modern technologies and to achieve equal levels of economic development.

CONTENTS

Chapter 1

Introduction

Agriculture is the process of producing food, feed, fiber and other desired products by the cultivation of certain plants and the raising of domesticated animals (livestock). The practice of agriculture is also known as farming, while scientists, inventors and others devoted to improving farming methods and implements are also said to be engaged in agriculture. More people in the world are involved in agriculture as their primary economic activity than in any other, yet it only accounts for four percent of the world's GDP.

Agriculture can refer to subsistence agriculture, the production of enough food to meet just the needs of the farmer/agriculturalist and his/her family. It may also refer to industrial agriculture, (often refered to as factory farming) long prevalent in "developed" nations and increasingly so elsewhere, which consists of obtaining financial income from the cultivation of land to yield produce, the commercial raising of animals (animal husbandry), or both.

Agriculture is also short for the study of the practice of agriculture—more formally known as agricultural science. Increasingly, in addition to food for humans and animal feeds, agriculture produces goods such as cut flowers, ornamental and nursery plants, timber or lumber, fertilizers, animal hides, leather, industrial chemicals (starch, sugar, ethanol, alcohols and plastics), fibers (cotton, wool, hemp, and flax), fuels (methane from biomass, biodiesel) and both legal and illegal drugs (biopharmaceuticals, tobacco, marijuana, opium, cocaine). Genetically engineered plants and animals produce specialty drugs.

In the Western world, the use of gene manipulation, better management of soil nutrients, and improved weed control have greatly increased yields per unit area. At the same time, the use of mechanization has decreased labor requirements. The developing world generally produce lower yields, having less of the latest science, capital, and technology base. Modern agriculture depends heavily on engineering and technology and on the biological and physical sciences. Irrigation, drainage, conservation and sanitary engineering, each of which is important in successful farming, are some of the fields requiring the specialized knowledge of agricultural engineers.

Plant breeding and genetics contribute immeasurably to farm productivity. Genetics has also made a science of livestock breeding. Hydroponics, a method of soilless gardening in which plants are grown in chemical nutrient solutions, may help meet the need for greater food production as the world's population increases. The packing, processing, and marketing of agricultural products are closely related activities also influenced by science. Methods of quick-freezing and dehydration have increased the markets for farm products.

Mechanization, the outstanding characteristic of late 19th- and 20th-century agriculture, has eased much of the backbreaking toil of the farmer. More significantly, mechanization has enormously increased farm efficiency and productivity. Animals, including horses, mules, oxen, camels, llamas, alpacas, and dogs; however, are still used to cultivate fields, harvest crops and transport farm products to markets in many parts of the world. Airplanes, helicopters, trucks and tractors are used in agriculture for seeding, spraying operations for insect and disease control, transporting perishable products, and fighting forest fires. Radio and television disseminate vital weather reports and other information such as market reports that concern farmers. Computers have become an essential tool for farm management.

A tractor ploughing an alfalfa fieldAccording to the National Academy of Engineering in the US, agricultural mechanization is one of the 20 greatest engineering achievements of the 20th century. In the early 1900s, it took one American farmer to produce food for 2.5 people, where today, due to engineering technology. Animal husbandry means breeding and raising animals for meat or to harvest animal products (like milk, eggs, or wool) on a continual basis.

In recent years some aspects of industrial intensive agriculture have been the subject of increasing discussion. The widening sphere of influence held by large seed and chemical companies, meat packers and food processors has been a source of concern both within the farming community and for the general public. There has been increased activity of some people against some farming practices, raising chickens for food being one example.

The patent protection given to companies that develop new types of seed using genetic engineering has allowed seed to be licensed to farmers in much the same way that computer software is licensed to users. This has changed the balance of power in favor of the seed companies, allowing them to dictate terms and conditions previously unheard of. Some argue these companies are guilty of biopiracy.

Soil conservation and nutrient management have been important concerns since the 1950s, with the best farmers taking a stewardship role with the land they operate. However, increasing contamination of waterways and wetlands by nutrients like nitrogen and phosphorus are of concern in many countries. Increasing consumer awareness of agricultural issues has led to the rise of community-supported agriculture, local food movement, slow food, and commercial organic farming, though these yet remain fledgling industries.

Very recently, genetic engineering has begun to be employed in some parts of the world to speed up the selection and breeding process. The most widely used modification is a herbicide resistance gene that allows plants to tolerate exposure to glyphosate, which is used to control weeds in the crop. A less frequently used but more controversial modification causes the plant to produce a toxin to reduce damage from insects.

Agricultural policy focuses on the goals and methods of agricultural production. At the policy level, common goals of agriculture include:

— *Food safety*: Ensuring that the food supply is free of contamination.

— *Food security*: Ensuring that the food supply meets the population's needs.

— *Food quality*: Ensuring that the food supply is of a consistent and known quality.

Chapter 2

Agriculture and Biodiversity

Agrobiodiversity refers primarily to genetic variability in cultivated plants and domesticated animals together with their progenitors and closely related wild species growing and evolving under natural conditions. Plants and animals gathered and hunted from the wild are also included in this term. Agrobiodiversity also provides employment and livelihood to a large section of the economically active population. On a global scale, it is noteworthy that nearly 2.5 billion people rely heavily on wild and traditionally cultivated plant species to meet their daily needs. Tribal women and men also serve as repositories of knowledge systems on the utilization of these plants.

While informal crop improvement by farm families is largely based on selection of superior types, scientific plant breeding draws heavily upon planned hybridisation among land races, locally adapted cultivars and wild relatives of crop plants. Such planned pyramiding of genes results in superior varieties combining high productivity with better quality traits, genetic resistance to diseases and pests, and greater tolerance to stress environments, including drought, salinity, water logging and extreme temperature regimes.

The commitment of the international community to ensure food and nutrition security to every child, woman and man on our planet has been reiterated at the following international conferences held since 1989:

— 1989 Bellagio Declaration : Overcoming Hunger in the 1990s

— 1989 Convention on the Rights of the Child

— 1990 World Summit for Children, New York

— 1991 UNICEF/WHO - Conference on Ending Hidden Hunger (Micronutrient Malnutrition)

— 1992 International Conference on Nutrition (ICN, 1992), Rome

— 1992 UN Conference on Environment & Development (UNCED, 1992), Rio de Janeiro

— 1993 World Conference on Human Rights (WCHR, 1993), Vienna

— 1994 International Conference on Population & Development, Cairo

— 1995 UN Conference on Social Development, Copenhagen

— 1995 Fourth World Conference on Women, Beijing.

— 1996 World Food Summit, Rome

— 1999 World Science Congress, Budapest

— 2000 Millennium Su nmit

Despite the high level political reaffirmation of the need to achieve speedily the goal of "Food for All", over 800 million human beings suffer now from hunger and malnutrition. Hidden and endemic hunger is now more due to the lack of adequate purchasing power at the household level than due to the lack of availability of food in the market. Poverty as well as factors such as environmental hygiene, sanitation and lack of safe drinking water are becoming major contributors to food insecurity at the level of both community and individual. Also, the earth's capacity to produce adequate food for the growing population is decreasing.

The increase in human population, now mostly confined to the developing countries, enhanced purchasing power and increasing urbanisation will lead to a large demand for food as well as more diversified food products in the coming millennium. It is important that both intensification and diversification of agriculture, particularly in the developing countries is based on sound ecological foundations essential for sustainable advances in crop and animal productivity. The emerging situation has led to the ringing of alarm bells by experts concerning the earth's capacity to produce adequate food to meet growing needs.

Among the many scientific components contributing to agricultural progress, the breeding of new varieties of plants, animals and micro-organisms has been especially important. The "Green Revolution" was triggered in Asia by high-yielding varieties of wheat and rice and hybrids of maize, sorghum and pearl millet. Success in breeding has, in turn, come from the intelligent use of genetic variability. However, this very process has led to increasing genetic homogeneity in cultivated varieties and hybrids.

Genetic homogeneity enhances genetic vulnerability to biotic and abiotic stresses, necessitating the increased use of pesticides. Indiscriminate use of chemical pesticides and mineral fertilizers will harm the soil, water, flora and fauna. In addition, it leads to new health hazards among the human population including the suppression of natural processes of immunity. Genetic variability provides the feedstock for plant and animal breeding and genetic engineering enterprises. While, on the one hand, human capacity to produce novel genetic combinations through recombinant DNA technology is increasing, the rate of loss of genes, species and habitats, rich in biodiversity, is also increasing.

The Convention on Biological Diversity (CBD) adopted at the UN Conference on Environment and Development (UNCED) at Rio de Janeiro in June 1992, is a clear index of the concern and commitment on the part of the international community for the conservation and sustainable and equitable utilisation of biodiversity. Biodiversity occurring in plants and, animals is now referred to as 'agrobiodiversity'.

The future of global food security depends on the success of the efforts in the conservation and enhancement of agrobiodiversity. The technical conference convened by the FAO in Leipzig, Germany, in June 1996, to consider issues relating to the conservation of genetic resources has urged all nations to implement a global plan of action for the conservation and sustainable utilisation of plant genetic resources for food and agriculture. Genetic engineering offers new opportunities for the development of effective Integrated Pest Management (IPM) and Integrated Nutrient Supply (INS) systems. However, the loss of every gene and species will limit the options for the future for genetic recombination.

COMPONENTS OF BIODIVERSITY CONSERVATION

From time immemorial, pastoralists and farmers have created a large number of intra-specific and inter-specific variants. Natural selection in the wild added more variants compared to those domesticated. These put together constituted the basis for food production through many centuries. This diversity is still being cultivated, tested, maintained, improved, stored as well as exchanged. Agricultural scientists depend on intra-specific variation. It is the interaction of each variety and breed with its ecological and socio-cultural setting that defines its ability to contribute to the overall goal of sustainable food production.

Traditional agriculture also ensured the survival of certain non-agricultural species due to management techniques as well as specialised adaptational requirements. Perennial maize, Zea diploperennis, is maintained even today because of slash and burn cultivation. Use of fire by pastoralists enabled the persistence of several grass species. Agriculture in marginal areas always supported diversity with sustained production. The 1960s witnessed the emergence of intensive agriculture aimed to maximise production.

Use of a single or a few improved varieties over large areas, use of chemical inputs and the lack of environmental as well as ecological concerns on the part of agriculturists led to a stage where production was maximised but at the cost of natural resources. Though this was a necessity at that time to feed the billions, time has come to reconcile some of the factors that can be reversed. As the Global Biodiversity Assessment puts it, " overwhelming evidence leads to the conclusion that modern commercial agriculture has had a direct negative impact on biodiversity at all levels: ecosystem, species and genetic, and on both natu-ral and domesticated diversity". What is now needed is a careful blend of modern agricultural techniques with traditional variants to assure a sustained food production for the decades to come.

Most of the domesticated plants and animals of today were identified centuries ago. Articles 8 (j), 10 (c) and (d) and 15 (7) of the CBD all recognise the role of indigenous and local communities in the conservation and improvement of genetic resources. Article 8(j) is particularly relevant for agricultural biological diversity because of the dominant role of farmers in the generation of intra-specific variability in crops and breeds and in the accumulation of specialised knowledge associated with these activities.

Since the 1980s FAO and the Commission on Genetic Resources for Food and Agriculture have served as a forum to promote the concept of "farmers' rights". Several countries, including India have incorporated this provision in their national laws. Article 8, in general, dealing with insitu conservation provided a framework for several countries like Sri Lanka, Brazil and Germany to have conservation activities of relevance to agrobiodiversity in protected areas. An extremely urgent task is the preservation and revitalisation of this priceless heritage.

Ecosystems and High Diversity

Certain ecosystems are considered to be high in agrobiodiversity because of the nature of the environment in which the species and communities have to exist, adopt and survive. Marginal environments like the coastal areas (with high salinity) and areas with deficient rain-fall tend to have more diversity. Special emphasis on conservation of agrobiodiversity in such areas would enhance the utilitarian value of the species (and the gene[s] they contain).

Ecosystems and Wilderness

The concept of wilderness is nonexistent in agricultural ecosystems. However, marginal areas tend to maintain wilderness and because of this such habitats support additional adaptational provisions facilitating the continued existence of traditional cultivars. This kind of wilderness also support enhanced evolution since the selection pressures are more.

Ecosystems are Representative

Agrobiodiversity in protected areas and certain regions like the Andean-eco region and the north-eastern Himalayan biosphere represent unique germplasm. Prioritisation of conservation must begin with such areas.

Ecosystems and Evolutionary Biological Processes

Though the association with key evolutionary processes is a problematic concept, little is known on the mechanisms of long-term evolution to enable singling out particular habitats and/or species. However, one specific example where a wild rice that evolved with the mangrove habitat, Porteresia coarctata Tateoka, which has the potential to transform rice cultivation in coastal areas stand out—emphasising the need to look at conservation of such habitats and ecosystems which support key biological adaptations.

ECOSYSTEM APPROACH TO CONSERVATION OF AGROBIODIVERSITY

The ecosystem approach is based on the application of appropriate scientific methodologies focussed on levels of biological organisation which encompass the essential processes and interactions amongst organisms and their environment. The ecosystem approach recognises that humans are an integral component of the ecosystem. Agro-ecosystems comprise a special category where people purposely selected crop plants and livestock animals to replace the native flora and fauna. These systems vary a great deal depending on the intensity of human intervention. Shifting cultivation, nomadic pastoralism, savannah mixed farming, rotational fallows and traditional home gardens involve low intensity of human intervention.

Improved pasture, mixed farming, multiple cropping and traditional horticulture involve a greater degree of human management. High-intensity management is characteristic of intensive cereal cropping and cereal-based farming systems, intensive livestock production, orchards and plantations. Increasing human intervention often leads to a reduction in the diversity of plant and vertebrate animal species. Continued harvesting and the unsustainable practices of plant, soil and pest management tend to lower the diversity of associated plants, invertebrates and microbes. Decreases in diversity of plant species may contribute to increased pest and disease problems.

Modern intensive agriculture utilises a strikingly narrow range of crop species and fewer crop varieties that have been bred for higher yield, including better response to inorganic fertilizers and incorporating built-in genetic resistance to selected diseases and pests. An ecosystem approach can be used in dealing with these issues. Both generic as well as ecosystem specific recommenda-tions flow from such analyses. As in the case of biodiversity as a whole, these issues have to be dealt with at the following three levels.

— The community (Habitat)

— The species

— The organism (genotype) / the Gene

The community: Natural communities represent the highest level of complexity with their diversity of species, individuals and genes in dynamic interaction, enhanced by natural and human selection in diverse environments.

The species: The species occupies a pivotal place in conservation programmes. According to the Global Biodiversity Assessment (GBA) report, about 1.9 million species have been described so far. The GBA also estimates that about 13-14 million species may exist, if we are able to identify and describe all invertebrates and micro-organisms. Two categories of species call for special attention. Domesticated species come under the first category. These are the ones which either directly or indirectly sustain humankind. Their wild relatives constitute valuable breeding material.

There is extensive intra-specific variability in many of the domesticated species due to variability in soil, hydrological status, micro-climate and human selection. The second category consists of species that are endangered in their habitats. Their preservation has become an issue of great urgency, since many of them may possess genes of inestimable value in agriculture, industry and medicine.

The Organism (Genotype) and/ or the Gene: The critical importance of individual genotypes and genes in plant and animal improvement is well known. Several improved rice varieties get their resistance to two of Asia's four main rice diseases from a single sample of wild rice, Oryza nivara. These varieties are now grown on more than 30 million hectares in Asia and have helped to double Indonesia's rice production. Similarly, wild cotton, Gossypium tomentosum, has provided a strain that does not produce the leaf nectar that attracts the insects, thus reducing the need for pesticides.

Wild genes, like wild species, are threatened both by habitat loss and loss of the on-farm conservation traditions of rural and tribal families. With advances in modern biotechnology, the identification, isolation and transfer of gene(s) from one species to another are possible. Conservation of specific genotypes, after a detailed character-based finger printing will help to accelerate the pace of plant breeding. This approach will be valuable, particularly for the incorporation of genes for tolerance to abiotic stresses.

Genetic Conservation Methods

While dealing with genetic conservation methods, it is important to distinguish the roles of in-situ and ex-situ conservation. In-situ conservation helps not only to preserve genetic variability, but also its enrichment through mutation, recombination and selection. Thus, it involves both conservation and continued evolution. In contrast, ex-situ conservation is more appropriately defined as preservation. Hence, it would be useful to refer to in-situ methods as conservation and ex-situ methods as preservation. An integrated conservation strategy will involve both insitu and ex-situ methods as well as in-vivo and in-vitro techniques.

An important feature of many plant communities is their resilience and ability to resume successional processes even after release from long-term disturbances such as grazing that may have been intense for decades. Under conditions of scarcity of financial and technical resources, a rational, scientifically based decision on priorities in conservation would have to rest on an assessment of the importance of the community, its current health and its likely resistance to further change.

Ex-situ conservation

Ex-situ conservation is an important and tested practice to support conservation. It primarily refers to preservation of genetic resources. While in ex-situ conservation evolution is largely stopped due to removal of material from its original habitat, in-situ conservation allows evolution to occur since the material are allowed to breed and recombine. Largely ex-situ preservation is carried out through gene banks.Following are the different types of genebanks

— An institutional genebank is set up to conserve only the germplasm which is used in the research programmes in its host institute.

— A national genebank is set up as a national plant genetic resource centre, maintaining many different germplasm samples of current and potential interest to people working in plant research nationally. Commonly, it contains germplasm which has been collected nationally. Also, it may be closely associated with a research programme or undertake its own research. A national genebank can be a collaborative venture between national institutes, or under the management of one institute which collaborates with other national institutes.

— A regional genebank is set up as a collaborative venture between a number of countries in the same geographical region to conserve the germplasm from that region and to support plant research.

— An International Agricultural Research Center collection is found in all the centers of the Consultative Group on International Agricultural Research (CGIAR) with a mandate for particular crops. Much of the germplasm is collected world-wide with international collaboration, and is conserved for the benefit of plant genetic resources activities world-wide.

Types of collections

— A base collection comprises a set of genetically different accessions of a given genepool conserved for the long term, ideally under long-term storage conditions. The international base collection of a genepool can be regarded as the sum total of all genetically different accessions conserved ex-situ in genebanks around the world. Base collections are not normally used as a routine distribution source.

— An active collection is a collection of germplasm for regeneration, multiplication, distribution, characterisation and evaluation. Ideally, germplasm in the active collection should be maintained in sufficient quantity to be available on request. Active collection germplasm is commonly duplicated in a base collection and is often stored under medium to long-term storage.

— A field collection or field genebank is a collection of living plants (eg fruit trees, glasshouse crops and perennial field crops) Germplasm which is difficult to maintain as seed can be kept in field collections.

— An in-vitro collection is a collection of germplasm kept as plant tissue. In some cases, the tissue is stored at very low temperatures such as under liquid nitrogen (cryopreservation).

A core collection attempts to store a large proportion of the genetic variation of a species within a manageable number of samples. Core collections are a mechanism to improve the accessibility and increase the use of collections. They are not a substitute for base and active collections. In the 1930s, international initiatives began in the 1940s, to collect and preserve crop plants under controlled environmental conditions. Such repositories, called gene banks, today form the important holdings of over 6 million accessions, being held in more than 1300 gene banks all over the world.

According to latest estimates 40% of the accessions are cereals, 15% are food legumes and 10% each of vegetables, roots and tubers, fruits and forages. Article 9 of CBD dealing with ex-situ conservation, emphasises the need for countries to support and enhance capacities of ex-situ conservation. But major steps need to be taken to strengthen such conservation efforts. These include:

— More than 50% of available collections lack appropriate passport data and hence are of little use. There must be appropriate efforts to fill the gaps of information

— Facilities for cheaper storage need to be developed

— The use of available collections must be encouraged

— Several ex-situ collections need to be segregated to assess their viability

— Policy measures on exchange and use of ex-situ collections need to be strengthened (in light of the patenting regimes over-shadowing their use).

Even though criticised for its biological status, ex-situ preservation is the most suitable way to preserve agrobiodiversity. Repatriation of the IRRI collection of over 7000 varieties back to India when the "Assam rice collection" was lost, the Seeds of Hope Programme in Rwanda and the LAMP project in Latin America are examples where ex-situ collections can help rejuvenate crop diversity when such diversity is lost. Issues relating to access and benefit sharing for ex-situ collection, especially those in CGIAR centers are points of discussion both with CBD and IUPGRFA processes. With the IUPGRFA taking final shape for implementation let us hope that these issues would be resolved.

In-situ conservation

Traditional agriculture has always been the dynamic interface between cultivated and wild species. It is based on cycles of careful selection and further introgression, over a period of time, suited appropriately for local environmental conditions. Such conservation efforts that is based on farm is called in-situ conservation. To mention the most important contribution of on farm conservation, it has been the diversity that it supports. This also has a strong undercurrent of traditional agricultural practices, knowledge systems and innovations.

Traditional farming have thus far contributed immensely to improve organised and commercial agriculture. Be it breaking the yield barrier or the resistance incorporation against pests and diseases, traditional cultivars have been of great use. But, in general, traditional cultivars do not yield well compared to modern varieties or high yielding varieties (HYVs). This has led to a lot of pressure on the farming communities in many parts of the world to discard traditional varieties and switch over to modern varieties.

The primary factor for this transition has been that of increased yield per hectare and quick crop. Such pressures on the farmers can be also attributed to policy considerations of different governments. These policies include those of giving subsidies for grow-ing HYVs, use of chemical fertilizers to maximise the yield and so on. CBD actively envisages a shift in such policies and encourages governments to adopt changes to let farmers continue traditional farming. While it is easy to propose the change it may be difficult to put it into practice owing to the increasing demand for more food. Here, it may be useful to consider the fact that transition from modern agriculture to sustainable agriculture need not be sudden and whole but may be slow and progressive.

To give an example, in countries like the Philippines there has been proven success with influencing farmers against the wide use of pesticides and instead towards the use of natural enemies of insect pests. Such practices coupled with transparent answers will probably be the best way to achieve the transition. Article 2 of the CBD takes the term "sustainable use" to mean "the use of components of biological diversity in a way and at a rate that does not lead to long term decline of biological diversity, thereby maintaining its potential to meet the needs and aspirations of present and future generations". In setting out its principles that guide the implementation of this objective,

Article 10 of the Convention provides five central elements that need elaboration here.

These are: the integration of considerations of biological resources into national decision making; adoption of measures that avoid or minimise adverse impacts on biological diversity; the protection and encouragement of customary uses of biological resources; support to local communities to implement remedial action in degraded areas where biological diversity has been reduced and encouraging cooperation between public and private sector in promoting the sustainable use of biological resources. Thus there is a clear indication of the interests of the Parties to the CBD in implementing the interests of the CBD. But what is required is more consolidation of approaches and collation of information on successes and failures. The Conference of Parties (CoP) to the CBD have categorically stated their immediate wish to consider actions relating to the conservation of agrobiodiversity. The following considerations may be helpful in helping both CBD and CoP in their efforts

Prioritising in Conservation

Global priority setting efforts are driven by several considerations. First, biodiversity is unevenly distributed, with some nations having more diversity than others, just as some ecosystems have more species than others. Biodiversity is useful not only for the local community but to the world at large when it comes to its use in agriculture, pharmaceuticals, human health and environment. Conservation of such resources require both national and international initiatives especially in terms of finances, expertise and information exchange besides prospecting. But, with biodiversity occurring at all levels and present everywhere, there is a need to prioritise the areas/species to be conserved mainly because of a lack of finances and trained manpower.

Hence there is a need to prioritise the initiatives with respect to biodiversity conservation. Taking agrobiodiversity into consideration, the need to conserve becomes more specific as we add value to the diversity for a specific purpose i.e. their use in improving global agriculture. For this to succeed there will be a necessity to prioritise conservation efforts on a regional level for global use. The principles for prioritising conservation include:

a) Linking priorities with clear objectives and goals

b) Laying down local, regional, national and international priorities

c) Evaluating the priorities

d) Understanding the needs and inputs of local communities and stake holders.

A crop like Teff grown widely in Ethiopia may not be important in all countries, but it is a key cereal in Ethiopia. Similarly, crops such as Chenopodium, Amaranths and Fagopyrum were in the past extremely important in the diet of the people in the mountain ecosystems such as the Himalayas. But these have tended to be replaced by wheat and rice in recent decades. A major reason for such a shift is the low yield potential of the crops grown earlier. Just as the introduction of the Norin dwarfing genes in wheat and Dee-gee Woo gen dwarfing gene in rice helped to raise the yield ceiling in wheat and rice, there is a need for launching a search for new genes for improving the yield potential of neglected and under-utilised plants. On the basis of this analysis, there is a need to accord priority attention to the following :

a) Saving endangered habitats, species and land races rich in agrobiodiversity.

b) Revival and revitalisation of the in-situ and on-farm conservation traditions of indigenous communities.

c) Supporting emergency conservation action in countries and regions affected by civil strife, ethnic conflicts or social disintegration.

d) Capacity building for launching integrated agrobiodiversity conserva-tion strategies involving appropriate combinations of in-situ and ex-situ measures.

e) Training biosystematics and conserva-tion professionals.

f) Helping nations to implement appropriate components of the Leipzig Plan of Action and to develop and implement a National Agrobiodiversity Programme for Sustainable Food and Livelihood Security.

Priorities for Action

(a) Promote research on traditional farming systems. There is a need to understand the dynamics and applications of traditional farming systems with a view to understanding their prudence in natural resource management, breeding for specific qualitative and quantitative characters and selection methods, besides learning about their integrated approaches to apportion their time, resources and energy.

(b) Understand reasons for the disappearance of such traditional farming systems: the reasons can be socio-economic or ecological. For example, the policies of government to encourage cropping of high yielding varieties, market forces, increasing influence of novel crop varieties and others may cause disappearance of traditional cultivation. Other reasons can be loss or fragmentation of habitats, increasing or decreasing availability of resources like water may necessitate local communities to abandon traditional cropping systems. Understanding reasons for these may help develop mitigating strategies.

(c) Enhance scientific understanding of the impact of conservation methods on large-scale commercial farms, including economic and ecological benefits. Concepts like Low-Input Sustainable Agriculture (LISA) are still lacking convincing evidence to support more research into influencing large-scale farmers to think about ecological benefits. Economics of traditional crop systems are also poorly understood. This situation needs more understanding, on a scientific basis, the impact of conservation methods on large-scale farming versus traditional farming.

(d) Promote, the concept of increasing the diversity of food basket. Nutritional requirements of local communities largely depend upon diversity of the food they consume. Promotion of in-situ conservation of diverse crops may help increase the diversity of their food baskets.

Management of in-situ conservation and encouraging such methods of conservation can be based on the following points:

(a) Integrating biodiversity impact assessment as a part of environmental impact assessment (EIA) procedures

(b) Encouraging traditional agriculture by providing suitable incentives

(c) Enhancing public and private sector cooperation

(d) Returning some nationalised resource systems, such as forests and wildlife to community based tenure rights

(e) Supporting establishment of a Farmers' Rights regime

(f) Developing equitable and fair benefit sharing methods to encourage use of traditional resources and increase participation of local communities in such efforts.

Arid and semi-arid ecosystem

Conversion of semi-arid or arid lands to agriculture (usually by providing irrigation) may increase local net primary productivity (the amount of plant material produced by photosynthesis per unit area, per unit time), but the effects on biodiversity are severe. When native plant communities are displaced, it is difficult to restore them even after cultivation is abandoned. Biological activity plays an important role in moderating the harsh environment of arid regions and drastic alterations in biodiversity may lead to the formation of alternative stable states of the system with stronger control by abiotic forces. Human alterations of biodiversity in desert regions have been mediated chiefly by the management of semi-arid and arid regions for livestock use and by the introduction of non-native plants.

Biodiversity which is moderately high in semi-arid regions declines with the increase in aridity beyond a critical level, more diversity of certain taxa adapted to these situations notwithstanding. Farming is usually highly diversified in this ecological region, both in terms of the number of crops and varieties within crops. Success in crop production depends largely on the right choice of crops and their varieties, keeping in view the soil moisture availability. Farmers, engaged in crop production and orchard keeping, also raise livestock such as cattle, buffalo, goat, sheep, pigs and camel, to supplement their income.

Under situations marked by high temperature and low humidity, soils are highly deficient in organic matter and nitrogen. Green manuring with leguminous crops and adopting suitable crop rotations and mixed farming are the key to maintaining remunerative yield levels. Application of irrigation for crop production may lead to the development of salinity problems because of poor management and inadequate drainage or use of poor quality water. Locally adapted farmers' traditional varieties are often highly tolerant to saline and drought conditions.

Growing of multipurpose shrubs and trees is common and many of these are life saving species during years of drought. Range lands and community managed pastures are a common feature. Grasses and forage species, grown traditionally, are highly drought tolerant. Arid and semi-arid regions are a reservoir of genes and co-adapted gene complexes that will be immensely useful for the development of new crops and superior varieties of existing crops, better suited to agriculture in marginal and stress environments.

A large number of cultivated plants, including cereals, pulses, fibre crops, oilseeds, vegetables, fruit plants, forage species and multipurpose shrubs/trees, have areas of their diversity in semi-arid areas of the world. Some prominent among them are sorghum, pearl millet, durum wheat, barley, pigeon pea, groundnut, chickpea, moth bean, cluster bean, green gram, black gram, horse gram, peas, flax, castor bean, tree cotton, upland cotton, sesame, cucumber, bottle gourd, melon, several forage grasses, alfalfa, pomegranate, date palm, mulberrry, jujube and numerous species of Salvadora, Acacia, Eucalyptus, Casuarina and Atriplex.

Promoting the use of under-utilized crops

At the global level, most of the world's major staple crops are served by substantial crop improvement programmes. International crop improvement efforts have delivered spectacular production increases, especially for crops such as wheat, maize and rice, at least in areas of relatively high potential. These crops received particular attention during the "Green Revolution" and continue to be the focus of major improvement efforts in both developed and developing countries. The private sector also tends to focus on the crops that cover either large acreages (maize, soybean, wheat, rice) or those that generate high per acre income (tomatoes, sugar beet etc.).

There is, though, a need for public sector funding of improvement programmes for other staple crops (eg. millets, cassava, sweet potato, plantains) where the commercial incentive is much less due to the lack of 'effective demand' by either producers and/or consumers. Far less resources are therefore invested in these food crops, which are sometimes described as "orphan crops". These crops however include many food and export crops which are important to many developing countries. Limited research and very few breeding programmes exist for many under-utilized species, despite the fact that they contribute to healthy and balanced diets, and offer farmers opportunities for agricultural diversification and additional income. Many countries have stressed the need for increased breeding and research on under-utilized species, and for domesticating many wild and semi-domesticated food plants.

There is also a potential for the development and marketing of new crops to contribute to the diversification of agricultural systems. This might also include the development of energy crops, for instance to deal with the growing fuel-wood shortages in many developing countries. The Global Plan of Action, through its activity 12, promotes the development and commercial application of under-utilized crops and species. The activity aims to:

a) identify under-used species;

b) develop sustainable management practices;

c) develop post-harvest and marketing methods; and

d) promote policies for the development and use of under-utilized species.

Agro-Forestry Conservation

Agriculture and forestry are often discussed as incompatible partners. The influences of pastoralists and farmers involved in shifting cultivation have contributed, atleast to some extent, to the destruction of forests. However, by the influence of natural balances of forest regeneration and revitalisation of soil fertility, forests and agriculture co-existed. Agro-forestry is not new in several parts of the world. Agro-forestry is seen as a good management of useful species, which allow conservation of a good part of animal and plant diversity.

Several studies have shown that biodiversity levels between natural forests, several agro-forests and monospecific plantations show the high potential of the original type of resource management system in conserving forest biodiversity in agricultural lands. Agriculture is seen as a useful buffer zone management technique outside protected areas.

Agriculture and Fisheries

Except for very few aquatic species, like the common carp, farming aquatic species is thousands of years behind crop and livestock farming. For long, it was felt that aquaculture need to be recognised as agriculture in that it uses land, water and nutrient resources as in the case of food crops. Aquaculture is considered in many parts of the world as a good source to low-cost food supplement, where farmers need to choose alternatives. Therefore, aquaculture can be considered as a part of integrated livestock and crop farming systems.

Aquaculture is one of the most important and rapidly growing food production sectors and is an important source of animal protein and income. Approximately 83% of production comes from the developing countries and the low-income food deficit countries account for 74% of the total aquaculture production. The integration of aquaculture into conventional agricultural systems is common in Asia. Un-consumed organic matter is recycled to increase production from fish ponds and subsequently the organic residues from ponds can be used as fertilizer for farmers' crops.

In several developing countries, crop rotation also involves culturing vegetables and fruits in fallow fish ponds. Rice fields that are flooded support a variety of fish species, increasing the productivity. Integrated and intensive farming systems are seen as a means to increase production, reduce risk of food shortages and minimise pollution. Thus agriculture and aquaculture are increasingly seen as activities that are mutually reinforcing each other, when considering food security issues.

Agriculture and Health

Modern requirements of organic foods and pharmaceutical products necessitates a sound knowledge of agriculture as a science-based technology. Organic foods are today finding a larger place in consumer's choice. Modern agricultural interventions like hydroponics are finding a greater demand not only due to the fact that they are safe and better but also because such agricultural practices also needs less space and can be highly intensive without having any effect on biodiversity, soil and water.

The growing need for organised cultivation of medicinal and aromatic plants for bioprospecting and the pharmaceutical industry requires an enormous understanding of the agronomic characteristics of such species. In addition, the harvesting, storage, and use of many of the medicinal plants needs inputs from several facets of agriculture like IPM, nutrient dynamics of soil etc. since the qualitative and quantitative characteristics of the medicinal and aromatic principles largely depend on the agri-management.

Agrobiodiversity Assessment and Agro-ecological Zoning

Agro-ecological zoning (AEZ) provides a framework integrating assessment, planning, management and monitoring of land resources. The AEZ concept involves a simplified representation of land with individual components of topography, landform, soil, climate, vegetation, land use etc. Antoine and Koohafkhan, explain the zoning process which involves a subdivision of the land on the basis of physical, biological and prevailing socioeconomic condition. These sub-divisions are called Resource Management Divisions (RMDs).

Such developments in databases and analytical tools in various land resource and use applications would have a strong implication on agriculture, especially in biosphere reserves where land use patterns have a strong relationship to biodiversity occurring within them. Changing land use patterns in areas—both under agriculture and those under non-agricultural practice—can be better assessed with respect to agro-biodiversity using the Resource Management Divisions (RMDs).

Agriculture and Land use Policies

Land tenures: Agriculture has always been an activity that weaves itself with several social and cultural practices and norms. Such norms and social rules impact the decisions of farmers in practising agriculture. One such important norms has been that of tenure rights in cultivated areas. Insecure land tenure rights affects the very basis of agriculture in most of the developing countries, where land holdings are becoming smaller and smaller. Such insecurities result in reduced farmer incentives to plan for long term productivity and management. Though improving tenure may not directly affect biodiversity, it would have indirect influences on protecting the benefits that a farmer can accrue from the diversity.

Insecurity of land tenure also influences intensifying agricultural productivity. This results in growth that is extensive rather than intensive. Activities such as agroforestry tend to receive less attention when the land tenure rights are insecure or short-term. It must also be noted that some times security in tenurial rights might also be a disincentive to biodiversity conservation as seen in the case of Brazilian Amazon, where conversion of natural ecosystems into agricultural lands proved disastrous.

Communal Rights: Community rights and / or common property resources (CPRs) from a major source of control over harvesting, use and management of natural resources in many parts of the world. Several countries, including Sri Lanka, are facing serious problems due to unsustainable resource utilisation by the communities and are finding it difficult to correct the damage. Public awareness and education can substantially correct the situation. However, the mismatch in government policies on property rights and common property rights are taking a toll on the biodiversity. The best example of this is the policy of the Government of Orissa, in India, to construct Jettys in Bay of Bengal to facilitate better navigation and eroding the CPRs of the communities living along the coastal areas in Orissa.

Laws: Rules and laws often tend to affect conservation. Brazilian tax and tenure laws that encouraged the clearing of Amazon forests is a good example of this. Also, extensive laws on felling and cutting of trees have often proved to be a disincentive for agro-forestry programmes in India.

Food Security and Trade

There are two broad options for achieving food security at the national level: the pursuit of food self-sufficiency or the pursuit of food self-reliance. Food self-sufficiency means the satisfaction of food needs as far as possible from domestic supplies with minimized dependence on trade. The concept of food self-reliance takes into account the possibilities of international trade. It implies maintaining a level of domestic production plus a capacity to import in order to meet the food needs of the population by exporting other products. Trade contributes to food security in a number of ways: through making up the difference between production and consumption needs; reducing supply variability; fostering economic growth; making more efficient use of world resources; and permitting global production to take place in those regions most suited to it.

Over time, global food security depends on maintaining and conserving the natural resource base for food production in both developed and developing countries. For any country, whether developing or developed, there is an unequivocal need for both access and the production of more and better food. This necessitates exchange of germplasm and improvement of a wide variety of crops. Increasing tendency to categorise crops as commercial and non-commercial crops threatens the very basis on which we recognise the importance of locally significant crops and those at global level. Greater emphasis should be placed on underutilised crops and lost crops. This will not only ensure conservation but also encourage a broader base for food production, especially at local and regional levels, minimising pressures on commercial food crops besides contributing to increasing micronutrition availability.

The Uruguay Round has been described as an important event in the evolution of agricultural policy. It paved the way to harmonise both national and international policies on agriculture for the first time. These advances are contained in a series of agreements and ministerial decisions and declarations, annexed to the Marrakesh Agreement. These are the Multilateral Agreements on Trade in Goods; General Agreement on Trade in Services; Agreement on Trade Related Intellectual Property Rights; Plurilateral Trade Agreements and Ministerial Declarations and Decisions. The Multilateral Agreement on Trade in Goods contains 13 individual agreements of which three are relevant to agriculture—agreement on agriculture, agreement on application of sanitary and phytosanitary measures and agreement on technical barriers to trade.

Agriculture and Biotechnology

Genetic engineering or biotechnology to assist crop breeding have been used for four broad goals: changing the product characteristics, improving the plant's resistance to pests and pathogens; increasing the yield and increasing the nutritional quality. Biotechnology is a broad term to describe various methodologies ranging from simple microbial process like fermentation to a complex genetic transformation work like transgenic production. For convenience let us concentrate on the genetic transformation technology which is receiving both support and criticism today. The FLAVRSAVR™ tomato is one of the first genetically engineered crop to be made available in the market. The fruit ripening characteristics of this variety were modified to increase shelf-life.

Biotechnology has also been used to change the proportion of fatty acids in Soybeans, modify the composition of canola oil, change the starch content of potatoes, and increase the production of pro-vitamin A in rice. Natural variability in plants to resist pest and disease attacks has been exploited for long by plant breeders. Biotechnology provides new tools to the breeder to expand his capacity. Unlike traditional breeding, biotechnology offers techniques to swap genes between species, between species and genera and also between plants and microbes. Thus gene transfers are independent of the gene's origin. Transferring the Bacillus thuringiensis (Bt) gene into crop plants like corn, rice, tomato is so far the most significant and popular use of biotechnology to confer resistance against lepidopteran pests in crop plants.

Other strategies to prevent insect damage include using genes of plant origin to produce proteins that retard insect growth (lectins, amylase inhibitors, protease inhibitors, etc). Use of Coat Protein Mediated Resistance (CP-MR) is popular to prevent viral attacks in plants. Apart from these successes to protect plants against pest and pathogen attack, several strategies seem to increase the potential crop yield, including the exploitation of hybrid vigour, delaying plant senescence, induction to flower earlier and increasing starch production. Use of cytoplasmic male sterility (CMS) was widely used, long before the age of biotechnology, to produce hybrid seed to increase yield potential. But strategies to exploit male sterility required biological manipulations that can be carried out using molecular biology tools. Successes of using such techniques are already on the scene.

Biotechnology also provided methods to increase the nutrient content of crop plants. Research into packaging a rice chimeric gene with a coding sequence for the highly nutritious hydrophilic protein like casein, under the control of prolamine regulatory and signal sequence, is found to increase the digestibility of prolamines in human intestine, resulting in the availability of vitamins. The possibility of enhancing Beta-carotene availability (a precursor of vitamin A) in rice is also exciting.

Incorporation of Ferritin gene into rice, transgenic bananas and rice with ability to produce vaccines invivo are some of the exciting developments in agricultural biotechnology. It is important that biotechnology based research also contributed enormously through other techniques like tissue culture, somatic hybridisation, wide hybridisation to increasing food production. However all the above possibilities exist because the gene(s) that can confer characters are available somewhere in the plant species. Without such variants it is impossible to practice any biotechnology. Genetic diversity thus form the basis for modern biotechnology and will continue to help food production in the years to come.

Impacts of Agriculture on Ecosystem Functions

The impacts of agriculture on ecosystem functions can be grouped into five areas: soil structure, nutrients and micro-organisms; water cycle, landscape complexity and linkages; and atmospheric properties. Agriculture affects soil structure and biota primarily through reduction of organic matter incorporated from above-ground activities and processes. Simplification of agriculture by removal of multi-storied vegetation results in exposure of soil and erosion. This affects soil invertebrates, microorganisms and soil insects which form an important component in decomposition and nutrient cycling.

Use of chemicals, as a part of agricultural practices, also affect these organisms. Soil composting, elimination of landscape features, lack of water infiltration, reduction of ground water recharge, changes in qualities of run-off water, affect the movement and quality of water needed for agriculture and its associated ecosystem functions. Intensification of agriculture also results in removal of several land areas like woodlands, hedges, fallow fields and smothens the wetlands, streams and ravines which all affect the biodiversity associated with such areas. Very often this also results in the loss of natural enemies of pests.

Research into effects of intensified agriculture have clearly shown how agriculture could influence the generation of greenhouse gases, impact carbon and nitrogen fixation. All these have impacts on climate change in the longrun. Methane released from flooded rice cultivation as well as ruminant production cause severe green house effect. Research into alternate cropping in such areas, like rice-wheat cropping, have shown how such deleterious effects of intensive agriculture can be reduced.

INTERNATIONAL INSTRUMENTS OF AGROBIODIVERSITY

Convention on Biological Diversity (CBD)

The objectives of this convention, to be pursued in accordance with its relevant provisions, are the conservation of biological diversity, the sustainable use of its components and the fair and equitable sharing of the benefits arising out of the utilisation of genetic resources. For the purpose of the convention, the term 'biological diversity' has been defined as 'the variability among living organisms from all sources including, inter alia, terrestrial, marine and other aquatic ecosystems and the ecological complexes of which they are part. This includes diversity within species, between species and ecosystems'.

The Convention on Biological Diversity has several articles relating to Technology Transfer (Article 16), determining how to establish a clearing house mechanism to promote and facilitate technical and scientific cooperation (Article 18.3), sustainable use of biodiversity (Article 10), sharing benefits derived from the use of biodiversity (Article 19.2) and involvement of and equitable sharing of benefits with indigenous and local communities (Article 8(j)). To achieve all this, it is necessary to a) identify components of biological diversity important for conservation and sustainable use, b) identify processes and categories of activities, which have or are likely to have significant, adverse impact on the conservation efforts and use, c) encourage use of biological resources that are compatible with sustainable use, and d) involve the local population to develop and implement remedial action in the degraded areas where biodiversity is either reduced or threatened.

This process has built-in mechanisms for introducing economically and socially relevant measures that can act as incentives for the conservation efforts of the local communities (Article 8(j)); exchange of information (Article 17); technical and scientific cooperation (Article 18); research and training (Article 12); public education (Article 13); access to and transfer of technology (Article 16); sustainbale use (Article 10) and identification of suitable financial resources (Article 20); are all important for arresting the loss of agrobiodiversity. Thus, the process of agrobiodiversity conservation and sustainable and equitable use will involve a blend of political will and action, professional skill and knowhow and peoples' concern and participation. Such a blend will have to be generated at the local, national, regional and global levels.

While the strength and capability of individual nations in the field of planning and implementing "Agrobiodiversity for Sustainable Food and Livelihood Security" programmes may vary, the collective strength of the international community for achieving this task is considerable. To facilitate this, Article 6(a) calls for development of national strategies for conservation and 6(b) for sectoral basis of implementation. Article 8 lays the basis for the in-situ conservation of biological diversity, which is set out as the most fundamental approach under the Convention. Article 9 relates to ex-situ conservation and is aimed at complementing in-situ conservation. Article 11 looks at implementation of incentive measures to farmers and thus encourage such practices.

International Undertaking on Plant Genetic Resources for Food and Agriculture (IUPGRFA)

The World Food Summit made a public commitment to end

hunger. Through the Plan of Action adopted by the Summit, governments, international organizations and all sectors of civil society are encouraged to join forces in an effort to ensure access at all times to the food required for a healthy active life for all the world's people. The Summit recognized the importance of plant genetic resources for food security and called on countries to implement the Global Plan of Action for the Conservation and Sustainable Use of Plant Genetic Resources for Food and Agriculture. This Global Plan of Action was adopted in June 1996 by 150 countries, at the International Technical Conference on Plant Genetic Resources, held in Liepzig, Germany. The International Technical Conference also adopted the "Leipzig Declaration" through which governments committed themselves to implement the Global Plan of Action. The Leipzig Declaration asserts that "our primary objective must be to enhance world food security through conserving and sustainability using plant genetic resources".

Global Plan of Action

The Plan aims to strengthen the links between conservation and utilization, through better information generation and management (Activities 1, 9, 17, 18), by improving the links between conservers and breeders through national plant genetic resources programmes and crop networks, and by greater investment in pre-breeding activities (Activity 10). Similarly, the Plan aims to promote not only the greater use of genetic diversity and resources, but also the strategic employment of genetic resources and practices which may lead to the maintenance of greater diversity in use (Activity 11). It also promotes an integrated approach to conservation, using both in-situ and ex-situ approaches and strengthening the links and complementarities between them.

The Plan focuses on action at the national level. particular attention is given to strengthening national plant genetic resources programmes which are regarded as pre-requisites to effective action. Whilst two activities of the Plan focus on these areas (Activities 15 and 19), action at the national level is stressed in many other activities. The importance of international collaboration is also recognised. The Plan gives particular importance to regional and sub-regional plant genetic resources networks (Activity 16).

The Plan promotes the full participation of farmers and local communities in planning and decision making processes relating to the conservation and use of PGRFA. The activity on national programmes (Activity 15) emphasises the need to involve all stakeholders, including farmers and local communities, with particular attention to women farmers. One activity area is devoted to on-farm conservation and improvement of PGRFA (Activity 2). Perhaps more significantly, the importance of farmer participation is integral to several other activities. Thus the importance of local PGRFA related knowledge (Activity 1) and of involving local communities in collecting (Activity 7), in situ conservation (Activity 4), evaluation (Activity 9), participatory plant breeding (Activity 11), management and development of under-utilized species (Activity 12) and seed distribution (Activity 13) is recognized. The training needs of farmers and local communities is also addressed (Activity 19). Also, the Global Plan of Action contains a special programme to support and restore traditional locally adapted farming systems in cases of war and natural disasters (Activity 3).

Finally, the Plan promotes complementarity between the public and private sectors, based on a recognition of the strengths of each. The Plan, by its very nature, focuses on those activities which need to be supported by public funds, especially at an international level. These activities include long term PGRFA conservation itself (Activities 2, 4 and 5 - 8), as well as other upstream, pre-competitive activities such as germplasm evaluation and pre-breeding, especially long term programmes to broaden the genetic base of breeders populations. Besides these "public goods", public sector support is also required to meet the needs of resource poor farmers for improved varieties and seeds which are suited to their needs, since they are often unable to express an effective market demand (Activities 2, 11 - 14).

IUPGRFA and the CBD

Article 15 of the Convention on Biological Diversity (CBD) deals with issues relating to access to genetic resources. Articles 15.2, 15.3, 15.4, and 15.5 deal with issues discussed under the IUPGRFA (International Undertaking on Plant Genetic Resources for Food and Agriculture) especially on access to PGRFA. Also, Article 15.3 deals with the issue of the access to germplasm pre-CBD. Similarly the International Undertaking (IU), Article 11, provides a number of options with respect to access.

One option provided for access in accordance with national legislation, and sharing out the benefits derived on a multilateral basis, according to a mechanism to be established by the Commission. This would apply to a list of genera, covering both in-situ and ex-situ material, as well as material collected before and after the entry of the CBD. The list could be based on importance for world food security and greater world-wide interdependence. There was wide agreement that this proposal might provide a useful starting point, although the disadvantages of limited inclusive lists were also stressed.

Another option was to bring an indicative list of genetic resources which directly or indirectly contribute to food security within the scope of the agreement, while allowing countries to include or exclude material according to agreed criteria. Various ways of developing lists were considered: (i) starting from a comprehensive list, and excluding those taxa on which agreement could not be reached, or (ii) beginning from a short agreed list and agreeing on further genera to be included. There was wide agreement that should a list be developed, provision should be made for countries voluntarily to designate additional materials under the agreement. Some countries in fact noted that they would be willing to designate all their plant genetic resources that are in the public domain. There was also agreement that any multilateral agreement should not preclude regional agreements with a more comprehensive scope. Article 15 (7) explicitly deals with the issues of benefit sharing which is also dealt with in IUPGRFA as 'farmers' rights'.

Intellectual Property Rights (IPRs) in Agriculture

(a) Traditional Forms of IPRs

Under the CBD, references to IPRs are to genetic resources and to conservation enhancing technologies. Two formal types of IPR protections are possible for protection of land races and other genetic resources. These are:

(i) *Patents* : Patents on genetic resources may be sought in the form of an entire organism or parts thereof, such as a group of genes, provided there is some human invention involved in it. Patenting of genes, excepting human genes, is not a legal problem and several countries allow such patents. Under the Trade Related Inteectual Property Rights (TRIPs) agreement micro-organisms may not be excluded form patent protection. Petty patents or utility patents are a weaker form of regular patent with less duration and limited royalty but is less expensive and easier to get. The crucible group, established by the CGIAR in 1995 advocated that genetic resources are best protected under utility patents.

(ii) *Plant Breeder Rights* : Plant Breeders' Rights (PBRs) as embodied in UPOV are a form of patent like protection especially for plants. PBRs apply to whole plant and are easy and inexpensive to acquire. PBR provides specific research exemption which provides access to protected materials. Varieties discovered in the wild are protectable with PBR, always they need to satisfy the homogeneity and stability requirement. PBR would apply to many of the needs for protecting genetic material in agriculture.

(b) Non-Traditional Form of IPRs

Many times traditional forms of IPRs may not be applicable to all forms of conservation, use and benefit sharing. Non-formal methods like the following may provide possible alternatives to protect the genetic resource as well as knowledge associated with it.

(i) Farmer's Rights - Farmer's Rights is the terms developed by FAO under the IUPGRFA. Farmer's Rights (sensu FAO) are described as rights arising from the past, present and future contributions of farmers in conserving, improving and making available plant genetic resources". However, the concept of Farmers Rights stays on even today as a moral obligation rather than to foster economic incentive to farmers.

(ii) Folklore - WIPO and UNESCO agreement on 'Model Provision for Nation Laws' on folklore is seen as parallel to protecting genetic material. The components of this agreement is expected to have several similarities to conservation of land races by farmers as compared to traditional folklore practices. But due to lack of helpful details on how to apply this agreement in implementing the protection of agricultural practices and knowledge, this still remains largely as a concept.

(iii) Code of Conduct - Beginning with FAO's code of conduct for plant germplasm collecting and transfer in 1993 several attempts are being made to develop codes of conduct on research, for ethics and collaboration. The CoP-4 decision (iv/8) to request countries develop non-legal mechanisms to address issues of access and benefit sharing gave a new life to the codes of conduct. Codes of conduct can be institution specific (like the code of conduct for the Royal Botanic Gardens, Kew, UK) or activity specific (like code of conduct for participatory plant breeding) or group specific (Code of Conduct for ethnobotanists). Operationalising such codes, however, seem to be slow and many lessons need to be learnt to effectively implement such codes of conduct in the field.

(iv) Apellations of Origin - With the classical controversy of using the name 'Basmati' for a variety of rice developed by Rice Tech, USA, the protection that countries are seeking on geographic apellation is increasing. Though formed in 1958 under the Lisbon Agreement and administered by WIPO, the Apellations of Origin is increasingly seen as a form of protection to material as 'geographical name of a country, region or locality, which serves to designate a product originating therein, the quality and characteristics of which are due exclusively or essentially to the geographical environment, including natural and human factors.

Convention on Combating Desertification

Desertification and drought are closely interlinked with issues such as loss of biodiversity, food security, population growth, poverty, climate change, water resources, deforestation and resource consumption. Desertification is also a social, economic as well as environmental problem and drought and land degradation can occur in dry climate zones. In relation to agriculture, the most valuable tool to conservation will be to preserve the knowledge of local farmers and indigenous people concerning dryland management and survival strategies. Their full involvement in the sustainable development of these drylands needs to be ensured.

The amounts of genetic diversity of plants and animals microbes in drylands are unique and important. Their adaptation, which evolved over hundreds of years, can be vital tools for sustainable use and development. Traditional techniques of water management, irrigation, soil nutrient replenishment, crop rotation, cropping pattern, agronomy are all important tools in conservation of agrobiodiversity in drylands. Future work, in collaboration with CBD UNFCCC, CCD and other related processes, are essential and vital. Under the CoP-4, SBSTTA4 directives countries need to come together to address the issue of biodiversity conservation in drylands on high priority basis and ensure that some of the poorest people in the world are fed nutritiously and sustainably.

Chapter 3

Biotechnology and Agriculture

Biotechnology is a powerful tool that can help us manage plants, animals, microbes and other life forms and processes including human health, for our benefits. The arrival of genetically modified crops is inevitable, regardless of national policies and protests by guardians of public interest. The cheapest is tissue culture, which is already being used widely in orchid and mushroom culture in the region. It is also used as a tool in plant breeding eg. in anther and embryo culture. At the other end of the scale, it costs about US$1 million to clone one single gene, according to one estimate.

Agricultural Growth and Poverty Alleviation

The impact of irrigation on agricultural intensification and increased crop yield has been very well documented, the marginal returns of irrigation compared to other factor inputs such as farm technology and other rural infrastructure development are still a controversial issue. Improved information and understanding of the scale of incremental benefit of irrigation and other factor inputs to agricultural growth and development and to poverty alleviation have large public policy implications on rural development policy. This is particularly more relevant in setting irrigation and agricultural investment and financing policies. This information is also important considering the recently increased global public policy priorities and thrusts on poverty reduction strategies.

In this chapter analyse the incremental impact (input specific effects) of irrigation and other factor inputs on growth of total factor productivity and its implications on poverty alleviation in India over the last two and a half decades. Total factor productivity is also called productivity of all inputs taken together, and it is different from conventionally understood productivity measures like crops yield, water productivity, or labour productivity.

The overall growth and technical change in the agricultural sector has large implications on expanding the economic base and poverty alleviation in a region. Past empirical studies have shown that ultimately growth in productivity of all factors (TFP) in agriculture is vital for alleviating rural poverty in developing countries. The actual impact of agricultural growth on poverty in fact varies by the nature, region, and time period selected for the studies. Though most of the previous studies have unequivocally demonstrated that agricultural productivity growth has a positive impact on reducing poverty in Asia, the existing literature on rural poverty has failed to examine the incremental impact of each of the factor inputs on agricultural productivity growth as well as their marginal impact on poverty alleviation, and rural income enhancement.

Several studies in India have illustrated that irrigation management has a profound role to play in the poverty alleviation process. Some of the recent aggregate level empirical studies in India have also shown that access to irrigation has a positive impact on poverty reduction. However, no straightforward relationship has been shown between irrigation and poverty alleviation; and the impact of irrigation on poverty alleviation depends on several other intermediate factors. Thus, an improved understanding of the structure of the impact of various factors and a quantification of marginal impacts of each of the factor inputs on poverty measures is important for developing efficient and effective policy instruments for poverty alleviation.

The empirical results show that there is no significant growth taking place in agriculture productivity when the level of all inputs use and their costs are taken together (in terms of economic and technical efficiency) over the last two decades. This productivity growth of all inputs is different from simple crop yield or labour productivity. The changing trends in irrigation and land and labor productivity are shown in Figure 1.

The regression results show that the marginal impact of irrigation on growth of productivity of all inputs is positive and significant with an elasticity of 0.32. This means that one percent increase in irrigated area has brought about an increase of about 0.32 percent in the productivity of all inputs (TFP) in India during 1970-94. This is very high when compared to the impact of other factors such as fertilisers, HYV, and road infrastructure where the elasticity varies from 0.04 to 0.09. The marginal impact of rural literacy on agricultural productivity is the largest among the variables selected for the analysis.

This large impact of rural education is possible considering the fact that agricultural productivity and rural development are directly related to the adoption of improved technology, selection of appropriate mix of crops and inputs, timely application of these inputs, and farmers' ability to effectively process market and price information and farm managerial decisions. The impact of road infrastructure is also positive and significant which perhaps captures the effects of market access in agricultural and rural development.

The rural poverty was unequivocally higher in a state with less extent of irrigation especially in the early 1970s. However, the relationship between rural poverty and irrigation has been decreasing in the recent past. The change in the relationship between irrigation and poverty across the states in India over the last two decades has been shown in Figure 2. It shows that

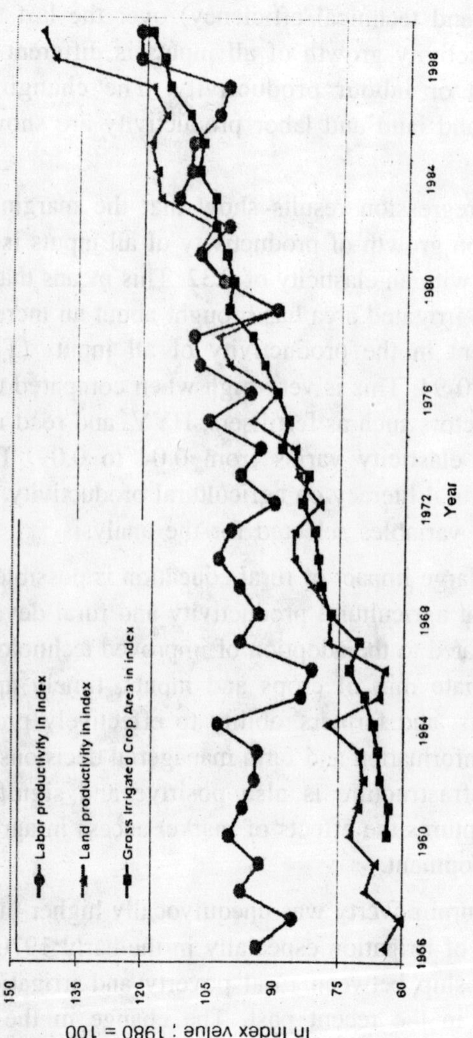

Figure 1: Changes in Land and Labour Productivity and Gross Irrigated Area, 1956-1989

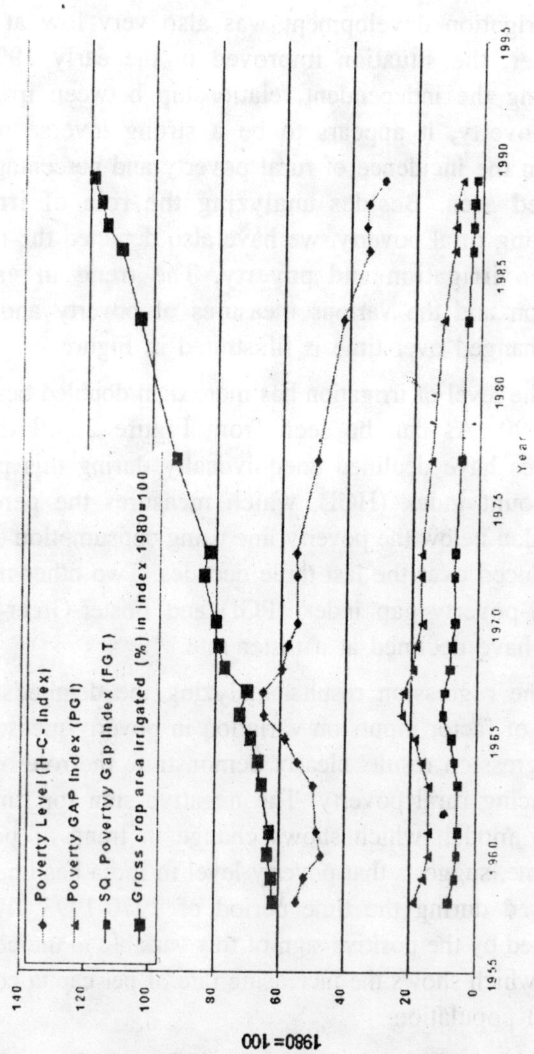

Figure 2: Relationship between Irrigation and poverty measures at all India level

the poverty level was very high in the early 1970s, with more than 60 percent of the rural population under poverty line (head-count ratio) in India.

Irrigation development was also very low at that time. However, the situation improved in the early 1990s. While analyzing the independent relationship between irrigation and rural poverty, it appears to be a strong inverse relationship between the incidence of rural poverty and percentage of gross irrigated area. Besides analyzing the role of irrigation in alleviating rural poverty, we have also depicted the relationship between irrigation and poverty. The trend in variation in irrigation and the various measures of poverty and how they have changed over time is illustrated in Figure 2.

The level of irrigation has more than doubled between 1960 and 1990. As can be seen from Figure 2; all the poverty measures have declined unequivocally during this period. The head count index (HCI), which measures the percentage of population below the poverty line using consumption expenditure has reduced over the last three decades. Two other measures of poverty-poverty gap index (PGI) and Foster-Greer-Thorbecke (FGT) have declined at a faster rate.

The regression results analyzing the detailedstructure of impact of factor inputs on variation in poverty measures (HCI). The regression results clearly demonstrate the role of irrigation in reducing rural poverty. The negative sign for time trend in poverty model, which shows change in trend of poverty rate over time, suggests that poverty level in India has unequivocally decreased during the time period of 1970-1993. This is also supported by the positive sign of this variable in the consumption model which shows the increasing rate of per capita consumption of rural population.

Irrigation has the strongest influence in explaining the reduction in poverty. Irrigation has even a larger marginal impact on reducing poverty than the impact of rural literacy. Likewise, increased HYV adoption and fertilizers use have also played a favourable role in reducing poverty in India, but their influence on poverty reduction is lower than the marginal incremental impact of irrigation and rural literacy. Unlike productivity growth, road infrastructure does not play any positive and favourable role in explaining the variation in rural poverty in India.

The future strategy of poverty reduction in rural India will largely depend on how efficiently the irrigation sector is managed and how effectively irrigation access is provided to a large number of farmers in the regions that have still not benefited from the green revolution of the 1970s and 1980s. In addition, the lowest income quintile of population would gain more from the irrigation development than the other upper income quintiles of the population just below the poverty line due to increased employment (wage rate increase as well employment security) and other feedback effects generated in the rural economy. Thus, increasing access to irrigation is in fact a pro-poor strategy to alleviate the severity and gravity of poverty in a region.

BIOTECHNOLOGY FOR POOR PEOPLE: A CASESTUDY OF MEKONG REGION IN THAILAND

However, the price tag is not always the real obstacle to poor farmers reaping benefits from genetically modified crops. The use of Bt cotton is said to have helped Chinese farmers reduce production costs by 14%. This is presumably the major reason Chinese farmers in 2000 planted half a million hectares of Bt cotton. In Thailand, protests by those claiming to be guardians of public interest, coupled with a number of high profile cases of export rejection on the ground of transgenic contamination, have led the government to restrict genetically modified crops to experimental fields only.

Nevertheless evidence of 'pirated' Bt cotton can still be found on farms in many provinces in the country, although the seed of 'bitter cotton', as the illicit Bt cotton is known locally, is sold at prices several times higher than those for regular cotton. Savings from the high cost of insecticide use, in terms of money and health, are the main reasons for its popularity among farmers. On the other hand, such illicit plantings are dangerous. The seed piracy is also doing farmers a disservice in the long run, as it acts as a strong disincentive for private investment in crop improvement.

The alarm about potential ecological risks from genetically modified crops has been raised by some scientists. Such concerns include resistance that can develop in major pest species and effects of substances that are products of transgenic modification (eg. Bt and other toxins) on non-target organisms through the whole trophic chain down to soil insects and micro-organisms. However, since fire was discovered, no technology has ever been absolutely witnout risk. It is important that these ecological risks are not considered in isolation. For example, the ecological risks of Bt cotton should be weighed against the ecological, human health and economic cost of pesticide use. By banning commercial planting of Bt cotton, Thailand has also missed an opportunity to monitor and weigh tradeoffs among all of these risks under local conditions.

Although genes are costly to identify and develop, once isolated they can be used in many different crops and ways. For example the Bt gene, which confers resistance to other insects as well as cotton bollworm, can also be used to protect against particular insect pests in corn, cabbage, rice and soybean. Once a gene is incorporated into a crop species, plant breeders without capacity for DNA technology may use the gene by the old fashioned methods of manual cross fertilisation and backcrossing. The more widely a gene is used, the more thinly can its research and development costs be spread. The licensing of Bt cotton from Monsanto in China could hold valuable lessons for other Mekong Region countries.

The case made against genetically modified crops is often based on a wide range of reasons associated with individual cases, and rarely on the actual problem specific to biotechnology. The safety record of the introduction of and experimentation with genetically modified crops in the region has indeed been miserable. The case made against Bt cotton in Thailand was based on accusations of a dereliction of duty on the part of the director general of the Department of Agriculture; a hard driving American culture of lobbying; the questionable ethics of politicians, public servants and scientists; and predatory business practices. These may indeed be important issues in the safe and equitable use of biotechnology in the region, in the way that bad driving is a major cause of traffic accidents throughout the Mekong Region.

To make these into indictments against biotechnology, however, is like saying cars must be banned altogether because we cannot get anyone to follow traffic laws, traffic police can be bribed, and the poor pedestrians and cyclists are more often killed than the rich in big expensive cars. On the other hand, some may consider nuclear energy to be a more appropriate analogy. While there are still so many uncertainties associated with risks from biotechnology, local regulatory as well as technical capacity would indeed be crucial for safe use of genetically modified crops in the Mekong Region.

Obviously, biotechnology cannot be expected to solve all of the problems faced by the Mekong Region's poor, but they offer some new options for those who still make a living from the land. Even among the various technologies for crop production, genetically modified crops are not always the right answer. The 'golden rice' (genetically engineered for β-carotene accumulation in the grain) may or may not turn out to be the best solution to prevent vitamin A deficiency.To combat widespread iron deficiency anaemia in the Mekong Region, grain iron in rice may be increased by genetic engineering or by using the genes conferring ability to accumulate twice to three times as much iron in the grain as standard varieties already found in local rice germplasm. Then again there are some who argue vigorously for other solutions, from fortifying food, teaching more nutrition in school, to educating more girls and, thus, mothers.

Biotechnology also provides many inexpensive tools for crop improvement and plant breeding. Facilities such as marker assisted selection (a gene is 'marked' so that it can be detected with simple equipment) can cut several years off the time, and hence the expense, of developing new varieties.The cost is within the budget of regional public research centres, medium SMEs (small and medium size enterprises) and regional universities in the Mekong Region. They are no more expensive to set up and run than the most basic soil and plant analysis laboratories that can commonly be found within the region. Capacity to make full and safe use of these tools, however, is another matter, and it does not depend on capacity in modern biotechnology alone.

From the mountains in northern Thailand to Yunnan we can find farmers who have not only improved their food security but become materially richer by going directly from subsistence production to high value cash crops, often with expensive imported seeds of crops such as cabbages, tomato and potato. Whether any of these crop varieties have been genetically modified is of less relevance than the fact that farmers even in the most remote places in the Mekong Region have chosen and are able to pay for and benefit from the product of commercial crop genetic improvement, in which biotechnology will inevitably and increasingly play an important role.

Commercially developed Bt cotton and other genetically modified crops offer at least one option through which poor farmers may choose to solve their food security problems. The obstacle to farmers taking this option is not the price of genes that their private owners will extract, but rather, public dissent. It would take a particularly inept seed company to set up prices that are beyond the reach of growers or to push a technology that will bankrupt them. If unconstrained by public dissent, Cambodia, Lao PDR, Myanmar and Vietnam can simply choose to follow China's example in adopting biotechnology if they so choose.

Genetically modified crops may also be transferred from China, as has already happened with hybrid rice in Vietnam. Public rejection of food made from genetically modified crops in importing countries in Asia and Europe is an important constraint for the whole region. Private seed companies, however, will not of their own volition pay attention to potential adverse impacts of genetically modified crops. There is also no profit motive for private investment in many of the crops and problems that are important to local farmers eg. rice and cassava.

Furthermore, there are questions regarding use and conservation of the region's valuable local plant genetic resources. Transfer of technology from outside the region, including the international agricultural centres, will always make a valuable contribution. However, for reasons that will be made clear later, the Mekong Region will have to increasingly rely on its own public research capacity. To fully exploit the potential of biotechnology and to forestall any adverse impacts on human health and the environment requires an agricultural research capacity far beyond modern biotechnology with its focus on genes and DNA.

The region contains the centres of diversity of many valuable crop species, including fruit (mango, banana, citrus), vegetables (the eggplants, gingers), orchids and timber species as well as sugar cane and the all important rice and its wild relatives. Many problems specific to these primary gene pools are overlooked or ignored by others in developed countries. The issue of genetic erosion of wild maize is important to Mexico, its native land, but is of no particular concern in the USA or Europe where it is not a native. Furthermore, many aspects of the problems of these native species require close monitoring.

Hybrid swarms, resulting from crosses between cultivated and wild rice, are common in the region. These are probably the primary source of diversity in local rice germplasm; new local varieties continue to be selected from these by farmers. They are also likely to give us clues to the dynamics of the local Oryza (the rice genus) gene pool. However, these hybrids closely resemble grassy weeds most of the time and can only be observed in the field for a few weeks each year. It requires someone who lives nearby to keep a close watch on them. Local insight and understanding are also often required to realise certain promises and opportunities. KDML105 is Thailand's premier variety for producing high quality aromatic rice, Thai Hom Mali. Rice grown from KDML 105, however, can fetch anything from between US$100 and $200 per ton. At present no scientific method exists to measure precisely what contributes to this price differential. Yet local rice mills pay farmers these different prices and everything else in-between everyday.

Specific aspects of local problems may constrain direct transfer of technical solutions from elsewhere. Solutions to common physical stresses such as drought, soil acidity or nutrient deficiency may be available from other parts of the world but applying them locally may require in-depth understanding of local conditions as well as solutions that go beyond simple 'adaptive on-farm' research. Technology for growing crops on acid soils on steep slopes that have been successful in South America is not ready-made for direct transfer to northern Vietnam, Lao PDR or Thailand (see below).

Global solutions may pose potential threats under certain local situations, as exemplified by the problem of the Thalassemia disease. This is a human genetic disease specific to the Mekong Region (along with only a few other places in the world) that results in various forms of anaemia. A study at Mahidol University estimated that 1% of the population in Thailand is affected, while carriers of defective genes account for 40% of the population.

Prevalence of the genes and the proportion of people affected are much higher in the north and northeast than in the whole of Thailand, and a similar concentration is likely to be found in other parts of the Mekong Region ie. certain parts of Yunnan, Lao PDR, northern Myanmar and Vietnam. The current Consultative Group for International Agricultural Research (CGIAR) system-wide effort to overcome iron deficiency anaemia poses a potential danger to thalassemics. Fortification of food with iron sulphate will certainly increase their iron overload.

Capacity to Use

The impact of enormous international and national efforts mounted in the last 10 years or so to improve biotechnology capacity in the developing world has been rather uneven in the Mekong Region. A minimal capacity is indicated in Cambodia, Myanmar and Lao PDR by the absence of these countries from regional collaborations such as the Asian Rice Biotechnology Network, Asian Maize Biotechnology Network, and so on. Vietnam participates in the Asian Rice Biotechnology Network. It has shown interest in biotechnology, but investment has so far been relatively limited. China's enthusiasm for biotechnology is matched by its investment in research funding through programmes such as the National Program on High Technology Development (known as the 863 programme) and the National Program on Basic Research in which agricultural biotechnology is a major component.

Capacity building has been served by special grants in the 863 programme to promote research by young scientists. National Key Laboratories have been established in the general areas of agricultural biotechnology, crop genetics and breeding in north, central and south China. The labs have been well equipped to conduct biotechnology and molecular biology research. Opportunities, facilities and support for biotechnology research are also provided by the Ministry of Agriculture, Ministry of Education and Chinese Academy of Sciences. The strength in biotechnology achieved by China can also be gauged by the number of internationally competitive grants Chinese scientists have won (eg. three or four of the new projects in the highly competitive Collaborative Crop Research Programme of the McKnight Foundation have been won by China). The list of crops on which biotechnology research has been conducted in China include rice, wheat, corn, cotton, tomato, potato, cucumber, papaya and tobacco.

Most of these are important in Yunnan, papaya and tobacco are indeed uniquely so. From this, and also research papers from Yunnan that have begun to appear in internationally refereed journals (Nature, Theoretical and Applied Genetics, Genetica etc.), it might be concluded that Yunnan, being carried along with the rest of China, is probably most advanced in terms of biotechnology capacity in the Mekong Region. Held up as the success story of biotechnology in a developing country is Thailand's happy experience with a diagnosis for viral pathogens of shrimps with a DNA probe. This has been credited with saving the country's shrimp farming at least US$1 billion since 1996. The country's commitment to biotechnology is evident in the establishment of the National Centre for Genetic Engineering and Biotechnology (Biotec) in 1991.

Biotec spends 30% of its research and development (R&D) budget on in-house research, and 70% is allocated to designated research projects conducted by universities and research institutions in the country. Biotec's current focuses are on shrimp farming; technology for the utilisation of cassava; rice (disease resistance and genome project); dairy (reproductive technology); genetic engineering (from disease resistant papaya and tomato, Bt cotton, to drought tolerant and aromatic rice); DNA finger printing (pathogens); and supporting trade with DNA diagnoses (to comply with various trade agreements and restrictions) ADDIN. Special funding has been provided from the national budget , and a major loan from the Asian Development Bank, to establish special research and PhD graduate programmes in biotechnology in major universities.

Thailand's research capacity received a major boost with the establishment of Thailand Research Fund (TRF) as the country's major research granting body, which began its operations in 1993. Two programmes of TRF specifically designed to strengthen the country's research capacity are the postdoctoral programme and the Royal Golden Jubilee (RGJ) PhD programme. The promise of biotechnology and molecular biology lies in the high degree of precision they bring to scientific inquiries and development and evaluation of solutions. Such precision is, however, wasted on complex agricultural problems unless both the target and solution are accurately defined (think of all those messy missiles hitting the wrong targets because of on-theground intelligence failures). It requires capacity in basic science directed at mechanistic explanation of the problem, as well as the solution, to make full use of biotechnology and to prevent possible adverse impacts.

Capacity in Basic Sciences

New genes and new varieties are not always the answer. Mapping and genomic studies of submergence tolerance genes may be very interesting, but improved rice varieties adapted to deepwater have had little impact, because their eating quality that was not up to local standards. Even when new genes are needed they would be only part of the solution, especially for these difficult environments. For acid soils in South America, crops (eg. wheat) were bred specifically for tolerance to aluminium toxicity. But the varieties were used in conjunction with acid tolerant legume cover crops in combination with a no-tillage practice that together protect the soil surface and draw nitrogen from the atmosphere.

Factors limiting crop production on the acid soils in the mountains of the Mekong Region, however, cannot be assumed to be the same as those in South America. They may also differ from place to place in the region, as do opportunities for a solution. Traditional rice varieties grown by Karen farmers in northern Thailand can yield reasonably well at a pH as low as four with a very low level of available essential nutrients such as phosphorus. Acid tolerance and phosphorus efficiency genes are apparently already available amongst the local rice germplasm of the Mekong Region's uplands. A research question that could help improve food security for people who live on some of these acid upland soils might be why the Karen rice that can often yield up to 4 ton/ha, sometimes yields only 0.8 ton/ha on neighbouring fields.

Currently lacking in the Mekong Region is capacity in the basic sciences needed to define such local problems, and to evaluate potential solutions. Recent progress in seemingly abstract understanding of boron nutrition in plants illustrates this second point. Soils low in boron are widespread from Yunnan to northern Myanmar, northern and northeastern Thailand and probably throughout northern Lao PDR and Vietnam. Nutrient management of crops produced on these soils would greatly benefit if we knew which of the major crop species (annuals and trees, mango, longan, lychee, durian, mangosteen, papaya, rambutan, jack fruit, tamarind etc as well as pioneer forest species), are able to recycle boron from their old leaves. This, in turn, rests on a basic understanding of boron mobility in the phloem of these local species.

The usefulness of such basic knowledge became evident when a group at the University of California, genetic modification to confer the ability to produce boron transporting sugar-alcohols has also enabled the plant to recycle boron accumulated in old tissues via the phloem and re-use it to build new leaves, flowers and fruit when supply from the soil runs out at critical times.

The situation with rice quality is another example of a bottleneck in crop improvement that biotechnology can help to correct. Rice quality used to be the preoccupation of just Thailand, but is now also worrying Vietnam and China sitting on their million ton mountains of unsold hybrid rice, and no doubt other Mekong Region countries too as their stock of unsold rice begins to grow. However, improvements such as better resistance to drought, diseases, insects, better efficiency in the use of plant nutrients and fertilisers, higher yield, more β-carotene, more iron and zinc etc, are wasted on rice that no one will buy.

The primary constraint preventing the fruit of biotechnology reaching rice farmers is our current inability to systematically put 'good quality' into new improved varieties. This, in turn, cannot be done with any degree of precision until we (a) know what characteristics constitute 'quality' and how this determines price differentials and how they are controlled, and (b) are able to manage the relevant genes and environmental factors accordingly. This requires a better understanding of basic ecophysiological and biochemical processes as well as of the genetic and environmental factors involved. Incidentally, anyone in Thailand proposing to work on this in collaboration with advanced labs overseas can expect a lot of flak about selling our national treasure and 'secrets'.

Another problem, especially for the less favourable agricultural environment, is caused by landscape scale and dynamics. Thus new genetically modified crops that are super tolerant to various abiotic or biotic stresses (eg. acidity, drought, too much water, nutrient deficiency, salinity, and pests, diseases and weeds) will always provide only part of the solution. In many instances new genetically modified crops with tolerant genes will make the problem worse in the long run. Weeds become resistant to herbicides, nutrient efficient varieties can deplete the soil even more efficiently, crop tolerance breaks down as pests and pathogens evolve, processes that lead to acidification or salinisation continue until the limit of the currently tolerant crops is exceeded.

Agricultural science graduates in the region are rarely provided with sufficient basic ecology to enable them to deal with such real world. Although subjects such as agronomy and soil science cover certain aspects of ecology, to most agricultural science faculties in the region, ecology is very much about birds, butterflies, lakes, meadows, pollution and wildlife; subjects only marginally related to agriculture. Since the success of integrated pest management, everyone now talks about integrated something-orother management. Unfortunately, basic understanding is rarely transferred along with the jargon by the time these have reached the national agricultural research system after a two weeks training course. An investment to provide the relevant aspects of ecology in undergraduate agricultural science education would cost a tiny fraction of the current biotechnology or information technology budget.

Scientific and technical capacity for agricultural development in the Mekong Region has come a long way since the pre-Green Revolution days before 1970. The progress has been sufficient for direct technology transfer from elsewhere, notably in the case of maize, rice and also the intensive livestock and feed industry. Unfortunately capacity for basic science related to agriculture that is needed to make the fullest use of biotechnology, as well as to forestall potential dangers, is still extremely limited.

The achievement of agricultural science becomes even smaller in proportion to the number and size of agricultural science faculties in the whole country. The focus on technology transfer may explain the direction taken by agricultural research in the past 30 years. While sufficient for the particular purpose, it also means that capacity for basic, mechanistic, explanatory research in agricultural sciences is much more limited than in the medical, biological and physical sciences. In the meanwhile it has become abundantly clear that farmers are much more effective in conducting adaptive research than young agricultural science graduates. Countless on-farm adaptive experiments are being conducted each season by farmers all over the region.

Link between Researchers, Farmers and Farming

Another critical weakness in agricultural research that will undermine biotechnology's effectiveness, is the link between agricultural research and farmers and farming in the region. This on-the-ground intelligence failure may be even more destructive than those wrongly targeted missiles. Farming systems, participatory research, participatory plant breeding, and numerous other ideas and methods aiming to improve researchers' recognition of the need to understand and respond to on-farm conditions have come and gone without much real, lasting effect. In part this is because the region's (much the same in many other places the world over) agricultural research has been entirely supply driven. There are limited mechanisms to assess research results for relevance to farmers' real needs as well as scientific and technical quality.

The team from Mahidol University and Chulalongkorn University that worked on the DNA probe for shrimp pathogens was made up of laboratory based scientists who probably had never heard of participatory research and similar jargons. They are, however, world-class biologists, who worked on a clearly defined problem in close collaboration with the shrimp industry and its farmers. Indeed, real differences can be made even by old-fashioned soil science and plant nutrition, backed up by dependable laboratory analyses, as demonstrated by a group at King Mongkut Institute of Technology in Thailand, working closely with durian growers. Famously high priced Montong durian orchards in Chantaburi have been successfully rescued from a threat of productivity decline due to fertiliser mismanagement that has led to soil acidification and nutritional imbalances.

Unfortunately such effective application of existing technical capacity to farmers' real problems is far from common in the region. This weakness not only makes targeting and fitting new technology to the production system difficult, it also means that insufficient attention is being paid to the potential adverse impacts of biotechnology and genetically modified crops. The danger this poses for the Mekong Region, especially from gene migration from genetically modified rice to the native Oryza germplasm (see below) makes one grateful to the proponents of the 'precautionary principle' whatever their reasons. Some form of quality control process in research management, driven by demand from farmers at some stage, would help guide research towards the production system. The region may, however, need to take lessons on the nuances of managing basic research directed at agriculture and balancing long term goals and short term applications, from places with long and successful experience in this, such as Australia, California and The Netherlands.

ADVERSE IMPACTS OF BIOTECHNOLOGY

Genetically modified crops are generally the focus of safety concerns related to biotechnology. The concerns focus on two areas: (a) food safety and (b) the environment, specifically genetic erosion.

Food Safety

The concern about food safety related to genetically modified crops is still largely based on fear of the unknown. Evidence of poisoning from food made from genetically modified crops is specific to certain compounds and groups of people and is actually quite rare. It would be most informative and useful to compare the frequency of cases of critical food poisoning from genetically modified crops and other forms of contamination, including 'natural' (from fungi, insects, and herbal 'medicines') as well as 'unnatural' ones (eg. methylated spirit, pesticides and prescription drugs).

The allergenic risks posed by GM food plants are in principle no greater than those posed by conventionally derived crops or by plants introduced from other areas of the world. Plant viral DNA sequences are commonly used in the construction of the genes inserted into GM plants, and concern has been expressed about this. One concern associated with GM foods is the possibility that genes introduced into GM plants might become incorporated into the consumer's genetic make-up.

This DNA is derived not only from the cells of food sources, but also from any contaminating microbes and viruses. Given the very long history of DNA consumption from a wide variety of sources, we conclude that such consumption poses no significant risk to human health, and that additional ingestion of GM DNA has no effect.

Environmental safety

There is a variety of concerns about the environmental safety of GM crops.

Local germplasm

Rice is a good case study for examining the potential impact of GM crops on local germplasm for several reasons: 1) it is the staple food for the region's entire population of 240 million; 2) the primary centre of diversity for rice, as well as the habitat of its several wild relatives, lie within the Mekong Region; and 3) it occupies a primary position in the region's farming, which still provides employment to the major part of its population. Equally importantly, so much national and international effort is now directed at producing new and better rice varieties through biotechnology and conventional breeding.

Transgenic rice varieties already in existence include those that that produce β-carotene, precursor of vitamin A; high iron rice; rice that has incorporated cry genes; proteinase inhibitors; lectin for insect protection; chitinase for sheath blight resistance; resistance to the herbicide glufosinate; and gene for the C_4

pathway. Many more can be expected in the not so distant future, including rice with genes for apomixis, and various other traits directed at enhancing quality characteristics as well as production efficiency.

Judging from the speed at which experimental lines, such as soybean in the 1980s and more recently Bt cotton in Thailand, 'escaped' to farmers' fields, new GM varieties of crops that show clear benefits to farmers will arrive on farms in the region sooner rather than later. Even before GM rice becomes widespread, it is imperative to assess and monitor potential risks to the native Oryzas.

Modern plant breeding is generally blamed for losses in genetic diversity from agricultural land simply by replacing diverse local populations with just a few uniform varieties. It has indeed been documented that since 1970 local rice varieties all over Asia have been replaced by high yielding varieties of the Green Revolution and hybrid rice, especially in Vietnam and China. However, local rice varieties can still be found on farm in the Mekong Region, even in Yunnan. There is no reason why the rate of displacement will now be faster with GM rice. A better understanding of why these local varieties are still used and perpetuated will not only help to preserve a valuable resource, but should also be useful in finding ways to improve productivity and local food security.

Modern biotechnology can make valuable contributions in precisely measuring the extent of diversity that exists, its geographic and ecological structure and dynamics. However, support from other sciences for basic understanding of plants and their environment (agronomy, biochemistry, botany, ecology, ecophysiology, evolution and genetics, as well as economics and the social reasons for germplasm conservation), will be essential. Many consider the erosion of the Oryza genetic system in the Mekong Region to have begun with releases of modern varieties which replaced numerous older varieties with only a few new ones. The modern varieties are products of modern plant breeding. They include promising local varieties that have been through pure line selection to make them genetically homogenous; they also include high yielding varieties and hybrid rice.

In Thailand high yielding varieties are still grown on a relatively limited scale. However, Thailand has taken to 'improving' local elite varieties since the 1950s. Through pure line selection, a local Thai elite variety from Bangkhla near Bangkok named Khao Dawk Mali became KDML105, which gave rise to RD15 and a glutinous rice RD6 through mutation breeding. These three varieties together account for 60% the country's main season planting, and more than 90% of the rice area in many provinces. The three are, unfortunately, genetically very close. In the early 1990s a blast epidemic decimated tens of thousand hectares of this presumably homogenous population of the KDML 105 stock in several provinces. This genetic homogenisation of local traditional cultivars continues in Thailand, and is now being repeated with local rice germplasm in Lao PDR (CGIAR website) and possibly Myanmar.

In addition to local varieties of cultivated rice, local wild rice populations have also been disappearing in many locations. Common wild rice of the Mekong Region countries include Oryza rufipogon, O. nivara, O. officinalis, O. ridelii and O. granulata. They form an important gene pool from which traits with potential use in rice production may be found and exploited. Some of these are likely to have been the source of genetic diversity in local rice germplasm. Habitat loss is a major cause of vanishing wild rice populations, but urbanisation is an inexorable force that the Mekong Region has somehow to live with.

At the same time, wild rice has become a serious weed in rice fields in many locations. Farmers and extension workers are looking for ways to eradicate it. Degradation of the wild rice gene pool will have serious implications for the future rice genetic system. There is an urgent need for a better understanding of 'diversity' in the Mekong Region's local germplasm of cultivated and wild rice in order that it may be better used and conserved. Biotechnology provides many valuable tools for the study of genetic diversity, from markers and micro-satellite analysis to molecular evolution.

Hybridisation and gene flow

The most common species of wild rice in the Mekong Region are O. rufipogon (the perennial form) and O. nivara (the annual form), which are ancestors of cultivated rice. Both of these are inter-fertile with O. sativa, the cultivated rice species. They cross readily with cultivated rice in the glasshouse, and there is strong evidence of hybridisation in the field. Wild rice is out-crossing while cultivated rice is self-fertilising. Hybrid swarms resulting from the cultivated and wild rice hybridisation contain large amounts of genetic variability. Although they are not recognised as such, the various forms of hybrids are commonly noted by farmers in the Chiang Mai valley. Weedy forms are sometimes called 'civet cat' rice, to be avoided when picking panicles for seed.

Discoveries of new forms growing in the field used to be the most common origin of new rice varieties in Chiang Mai and other rice growing areas in northern Thailand, and presumably in other places in the Mekong Region. A population of 'wild rice' was collected by Oka's group in Sanpatong, just south of Chiang Mai, some time before 1961. Based on characteristics such as awnedness, panicle size (length and number of spikelets), spikelet width and seed shedding, it was found to be very heterozygous, containing genotypes that ranged from wild to cultivated types. Some were indistinguishable from cultivated types by external appearance. A research team from Chiang Mai University has since found other similarly genetically diverse populations in other parts of Thailand.

Gene flow from introduced germplasm into wild populations would have been going on ever since rice varieties were moved around the region (eg. with the different waves of migrations from southern China into Myanmar, Thailand, Lao PDR and Vietnam and various maritime, river and over-land trade traffics that plied the whole of the region, including mountainous areas long before colonial time). The introduction of GM rice and the advent of modern biotechnology, however, add some new and potentially dangerous elements to the local rice-wild rice genetic system.

This involves the process of gene 'transformation' ie. introduction of the 'exotic' or 'trans' genes from other species. The Bt gene that confers tolerance to insect damage comes from a bacterium called Bacillus thuringiensis. Other GM crops often contain genes transferred from micro-organisms, including viruses. A potential danger lies in the possibility of genetic interactions of these transgenes with other major genetic systems, from the crop species and its wild relatives to pests, pathogens and weeds and beneficial insects, micro-organisms such as nitrogen fixing bacteria and mycorrhizal fungi, and various other life forms down the trophic chain. The simplest scenario for such genetic interactions would be an 'escape' of a gene for resistance to specific herbicides from GM rice into wild rice.

The biotechnology method of embryo culture and embryo rescue has also enabled 'wide-crosses' (ie. hybridisation between rice and other species of the Oryza genus) to be made. Wild rice is increasingly used as a source of disease resistance and other useful genes (eg. cytoplasmic male sterility for hybrid rice). For example, Pathumthani 1, one of Thailand's new non-photosensitive, aromatic rices, contains some wild rice genes.The genetic barrier between cultivated rice and its wild relatives is likely to be much reduced in cultivated rice varieties that have incorporated wild rice genes.

A recent study by the USA National Research Council concluded that gene migration between the transgenic crop and wild populations of squash in the United States could pose an environmental risk. This risk is associated with introgression of the transgenes into wild populations. In this native germplasm, the researchers found weak but clear evidence of p-35S, a promoter from the cauliflower mosaic virus, widely used in transgenic crops, presence of the nopaline synthase terminator sequence (T-NOS) from Agrobacterium tumefasciens (another sign of transgenic contamination) and cry-1A gene of Bacillus thuringiensis (one of the Bt genes).

There is as yet no general rule to predict what the outcome will be when transgenes from GM crops migrate into local germplasm, wild or cultivated. Each genetic system is different, so must be examined on a case by case basis. An important lesson for the Mekong Region behind the above message from Mexico is the major role of local research in the first discovery of transgenic contamination and subsequent research for verification and confirmation by Mexico's own public research organisation and in collaboration with a public research institution in the USA, the University of California at Berkeley.

The urgency at which the Mekong Region needs to examine possible implications of gene flow in rice is indicated by the evidence of hybridisation between wild and a high yielding variety of cultivated rice in Thailand. Gene migration is suggested by the appearance of many domesticated traits (eg. prolific reproductive capacity, lower dormancy, husk and pericarp colour, grain shape and size, grain quality, panicle type, awnlessness, shattering resistance and photoperiod response) in the wild population and wild traits (awns, grain type and colour, shattering etc.) in the cultivated population. Capacity in modern biotechnology is essential to confirm if gene migration has really taken place, and to further examine possibilities of how GM rice would affect the Mekong Region's local Oryza genepool.

GM crops are already accepted as a matter of national policy in China. The forces of globalisation will make inevitable the arrival of GM crops in the remainder of the Mekong Region, regardless of whether or not the region has all the necessary technical capacity to handle them. However, to fully utilise biotechnology and forestall any adverse impact of GM crops the region will need to build up a technical capacity that goes far beyond the molecular genetics, genomics and DNA related science.

INTELLECTUAL AND TANGIBLE PROPERTY RIGHTS

In addition to technical problems, another stumbling block for the use of biotechnology in the Mekong Region as elsewhere is the issue of intellectual and tangible property rights.

International and National Legal Framework

The Agreement on Trade Related Aspects of Intellectual Property Rights (TRIPS)has obliged developing countries to enact laws to confer ownership rights to intellectual and tangible property which are products of creativity, invention, and know-how (intellectual property) as well as biological materials and devices (tangible property). Means of protecting such property that have been applied to plant germplasm include patents, trade secrets, trademarks and geographic marks and appellation origin. The last one allows legal rights to make and market certain kinds of products to certain geographical regions only, eg. Champagne from the region by that name in France, and similarly for Bordeaux, Burgundy, Cognac, Scotch, and so on. This is currently applied to wine and spirits only, but there are signs that agreements are being reached in the World Trade Organisation (WTO) to extend the application to food and other agricultural products as well.

On the plus side for the Mekong Region, it will mean that no one else will be allowed to export Thai rice or Yunnan ham. But there will also be a down side, in that the Mekong Region will no longer be able to export Basmati and Japanese rice, or other products claimed by other geographical regions. The UPOV (International Union for the Protection of New Varieties of Plants) Convention of 1961 established the principle that an improved variety can be legally owned by the breeder. The Convention is concerned with protecting the results of conventional plant breeding, so is generally believed to have the full weight of the seed industry behind it. The Exceptions of the Breeder's Right (Article 15) allows plant breeders to use without restriction protected varieties in the production of new varieties.

In its 1991 revision the UPOV Convention was brought in line with contemporary technological developments. In particular, it extended protection to 'essentially derived varieties' in an attempt to strengthen protection for plant breeders who initially develop a new and distinct variety against others who merely make derivations from the initial work. Article 14, 5 (c) says that "Essentially derived varieties may be obtained for example by the selection of a natural or induced mutant, or of a somaclonal variant, the selection of a variant individual from plants of the initial variety, backcrossing or transformation by genetic engineering".

Many developed countries have enacted legislation to protect plant breeders' rights in new varieties which complies with the UPOV Convention. For example, Australia's Plant Variety Right Act 1987 broadly followed the criteria for protection agreed under UPOV, and its Plant Breeders' Rights Act 1994 reflects the 1991 revision and ensures that Australian law complies with UPOV. However, no developing countries has yet joined the UPOV. The process to institutionalise intellectual and tangible property rights protection for plant varieties in the Mekong Region, along with other developing countries, is being strongly influenced by international developments.

In the meanwhile, many international rules and agreements themselves are still being argued. For example, an International Convention on Plant Genetic Resources for Food and Agriculture is currently being negotiated at the Food and Agriculture Organisation (FAO) in Rome. This treaty is expected to establish rules under which research germplasm can move within the global scientific community in order to secure crop improvement and food supplies in the face of global warming and rapidly-changing agricultural environments and regulations. Certain developed countries are, however, reported to be pushing to renegotiate (ie. refusing to share the financial benefits achieved through a free flow of scientific germplasm with the countries that provide the seeds) and to maintain the de facto right to patent any germplasm received after undertaking the simplest process selection.

Chapter 4

Rural Development through Agriculture

Agriculture and the Millennium Development Goals

Agriculture can make significant contributions to attaining the Millennium Development Goals (MDGs). It is the sector from which most of the rural poor in developing countries derive their income, and both rural and urban people obtain most of their food, which is produced largely by women. As agriculture depends heavily on the natural resource base, it influences environmental sustainability. Agriculture is also closely linked to human health and education.

An estimated 1.2 billion people are absolutely poor, living on less than US$1 per day; nearly twice that number live on less than US$2 per day. Currently, about 800 million people go hungry each day. Approximately 75 percent of the absolute poor in developing countries live in rural areas, where they depend mostly on agriculture for their livelihoods. Thus, reducing poverty in rural areas, and hunger in both rural and urban areas, will depend heavily on the sustainable development of agriculture.

Through efforts in the sector, income of the ruralpoor must increase rapidly, and food production in the developing world must more than double over the next twenty years to keep up with population growth. To achieve these goals, the sector must promote pro-poor economic growth at rates at least as fast as population growth rates. This, in turn, will require raising agricultural productivity, integrating agriculture into local and international markets effectively, and creating productive on- and off-farm employment.

Women are responsible for half of the world's food production and between 60 percent and 80 percent of the food in most developing countries. Not only are women the mainstay of the agricultural food sector, labour force, and food systems, they are also largely responsible for post-harvest activities. Their specialized knowledge about genetic resources also makes them essential custodians of biodiversity for food and agriculture. However, women's fundamental contribution is continually under-appreciated and under-supported, and is often adversely affected by prevailing economic policies and other development conditions. These circumstances must be reversed: sustainable rural development through agriculture cannot be achieved without the full participation of women.

The natural resource base of suitable land, water, forests, and biodiversity largely determines the potential of agriculture. These resource endowments have a major influence on human activity in agriculture, and in turn, are affected by them. Historically, agriculture responded only to the need for food. Much later, it sought to respond to poverty-reduction mandates as well. Now it seeks to simultaneously help meet the triple objectives of poverty reduction, food security, and environmental sustainability.

Most of the land suitable for agriculture is already in production. Therefore, meeting current and future food requirements will require rapid increases in productivity; otherwise, an undesirable expansion onto fragile and marginal lands will result. There is widespread concern that deforestation and land degradation are severely diminishing the potential of ecosystems. The main causes of these conditions go well beyond agriculture; however, agriculture does play a role: when policies are inappropriate, unsustainable agricultural practices are used and property rights are insecure.

Biodiversity supports the production of an ecosystem's goods and services essential for life as well as for many cultural values. Improving crops, livestock and feeds; increasing soil fertility; and controlling pests and diseases often depend on these resources; however, increasing population pressure, deforestation, and unsustainable agricultural practices are contributing to degradation of these "life insurance policies."

Good health and education are two prerequisites for sustainable development, and agriculture contributes to both-in positive, as well as negative, ways. Adequate nutrition is indispensable to attaining good health. Though insufficient by itself, an adequate supply of food is a key determinant of adequate nutrition. This factor alone can drastically reduce malnutrition in adults and children, and increase birth weights of newborns. By improving incomes and nutrition, gains in agricultural productivity can help break the cycle of passing malnutrition from one generation to the next. In addition, it is often the savings from agriculture that provide the means to meet expenses relating to educating children.

Agricultural practices, however, can have negative effects on human health and education. For example, overexposing adults and children to dangerous chemicals and harmful forms of child labour in both family and commercial settings are significant problems. In addition to exposure to dangerous chemicals, children may suffer long working hours, lack of access to education, very low or no pay, and injury due to heavy loads and dangerous machinery. If children must work to support themselves or their families, they should be assisted with programs that reduce the physical risks they face and provide leisure time, flexible schooling, and fair pay. Agriculture and health are also related to efforts to combat HIV/AIDS, malaria, and other diseases. Poor people and farming communities have been particularly hard hit by HIV/AIDS: about 60 percent of HIV-positive sub-Saharan Africans are women. Given women's pre-eminent role in food production and preparation, this fact could exacerbate food insecurity in the region.

Linked to other Areas

Agriculture is functionally linked to many other areas. The Agency's agricultural programming will, therefore, reflect these linkages to achieve integrated development and ensure that its total development effort is greater than the sum of its parts. Agricultural programming is linked to the following areas, among others:

Private sector development

Small farmers in rural areas often comprise the largest segment of the private sector in developing countries. The full potential of these farmers and farms is often not realized due to poor policies, inadequate markets and other infrastructure, and generally weak institutions. Creating the enabling environment in which agriculture can perform is crucial. For example, well-functioning agricultural markets can underpin a rural economy and help promote rural enterprises in agro-processing and the provision of rural services.

Water

Water is an indispensable resource for agriculture and has played a pivotal role in the development of the sector, but it is also scarce and unevenly distributed both regionally and among certain marginalized populations. Agriculture is the largest user of water, accounting for more than 70 percent of total freshwater withdrawals globally and between 87 percent and 95 percent in developing countries. Current water use by agriculture may not be sustainable because of both scarcity and competition for use from other sectors such as human consumption, health, sanitation, and industry. As a result, many innovations to improve water use efficiency are being tried, and others such as more water efficient crops are needed.

Forestry

Forests have historically provided shelter, food, fuel, medicines, and building materials. More recently, they have become sources for new goods and services such as pharmaceuticals, raw materials, recreation, and carbon sequestration. However, forests now cover only 24 percent of the world's land surface, and a net loss is occurring in developing countries. There are no simple answers to deforestation, but developing sustainable agricultural systems will help ensure that forests continue to provide both traditional and new goods and services.

Oceans and fisheries

The oceans offer a development option to a major portion of the 660 million inhabitants of the least-developed countries. Like dairying, the sector provides an all-year harvest and income stream, which contributes to the social and economic welfare of large cross-sections of rural and urban populations. Since staple foods are often protein-deficient, poor people can improve their diet by adding fish, which are rich in protein. Approximately 50 million women are employed in this sector. Access to resources within the 200-mile exclusion zones of coastal states has brought new opportunities, and valuable social and economic assets under their control.

Pursuing an Integrated Approach

The internal complexities, as well as the external linkages, of agriculture must be simultaneously managed to attain development that is integrated, equitable, and sustainable.

Sustainable Livelihoods focuses on activities that promote sustainable human communities. The approach begins with peoples' assets and capabilities, and seeks to build on them. A livelihood is said to be sustainable if it can adapt to stresses and shocks, maintain and enhance its capabilities and assets, and at best, enhance opportunities for the next generation. This approach recognizes that the root of all human development and economic growth is livelihoods-not jobs per se, but the wide, infinitely diverse range of activities people engage in to make their living-together with assets or entitlements they own or can access. Hence, it integrates considerations of income generation; the production of sufficient, nutritious food; women's empowerment; and environmental management.

Ecosystems Health is a way of thinking about human development that focuses on the systemic ecological and social contexts in which human activities occur, and that make them sustainable or not. A healthy ecosystem maintains itself without major human intervention, changes and adapts over time, and provides the services that sustain human communities. This approach, therefore, provides a broad framework to help identify both constraints and opportunities for those activities. It can help to identify agricultural policies and practices, for instance, and the livelihoods associated with them that increase food production without disempowering women and undermining ecological integrity.

Developing countries striving to transform their rural economies through agricultural development are confronted with many challenges that are both national and international in scope. At the national level, careful participatory planning, appropriate policies, and sound implementation strategies are prerequisites. Additionally, pursuing new scientific opportunities made possible by the "gene revolution" and advances in biotechnology will require careful balancing of the opportunities and risks.

At the international level, developing countries must effectively participate in global governance mechanisms relating to agriculture in the form of conventions and protocols, and trade. Agreements, such as those of the World Trade Organization, lock developing countries in for the long term and have direct impacts on local economies. Most developing countries are often not in a strong bargaining position to negotiate such agreements. Additionally, they are further disadvantaged by having insufficient technical capacity to participate effectively in these negotiations.

Developing countries require capacity building to participate in the development of international frameworks, to develop national policies reflecting domestic priorities and international requirements, and to implement these policies and meet international obligations. All together, these capacities must enable developing countries to sustainably manage their rural economies and the natural resource endowments on which they are based.

Programming priorities:

— Supporting sector assessment, domestic policy formulation, and strategic planning.

— Assisting developing countries to compete both regionally and in the global marketplace through enhanced capacity to formulate and implement trade policies, and to develop infrastructure and overcome trade barriers, e.g. sanitary and phytosanitary requirements.

— Building the capacity to respond to the opportunities and risks of biotechnology.

— Increasing national capacity to undertake gender analysis, manage natural resources, and respond to agriculture-related conventions and protocols.

Creating and new knowledge for development

Knowledge-both indigenous and new-has been pivotal to agriculture's development contributions. Such knowledge has historically supported food security objectives, but must now also contribute to poverty eradication and environmental sustainability as well. Attaining the potential contributions of agriculture to the MDGs will depend on continued creation and use of new, as well as existing, agricultural knowledge at an accelerated rate. Much remains to be done about old problems such as land degradation, plant and animal pests, and disease control, even as new challenges such as climate change, water scarcity, and pressure on biodiversity are emerging. Lessons from past experiences with respect to second-generation technological problems of irrigation, cultivation practices, and excessive inorganic chemical use, must also be learned and addressed.

New science in the areas of genomics and biotechnology, and biological control of diseases can potentially improve crop and livestock adaptation to environmental stress, including climate change, which will, in turn, improve yields and conserve natural resources. These avenues must be approached in a balanced way, carefully weighing benefits and risks. Contemporary research and transfer organizations have not been as effective as they might have been. Improving them will require greater attention to the role of women in agriculture and to the appropriateness of innovations, which are to be designed, tested, and transferred using participatory approaches.

Programming priorities:

- — Strengthening national, regional, and international agricultural research and transfer capabilities.
- — Improving crop and livestock adaptation to stress and enhancing the efficiency of natural resources utilization.
- — Increasing the food and feed value of staple crops of the poor.

Enhancing food security through agricultural productivity

Agriculture in developing countries is increasingly moving away from a subsistence orientation and government dominance to commercialization. There are opportunities to accelerate this process in such a way that producers, particularly women, who produce a dominant share of the world's food, become equal partners in the development process and share the benefits. However, the rural poor, particularly women, own or have secured access to few assets they can use to escape poverty. Secured access to land, for example, is often a binding constraint, and the poor are often left to cultivate the marginal areas. The productivity of their two main assets-land and labour-is very low, but can be significantly improved through intensification and diversification of their production systems. Potential strategies include matching production with natural resource endowments, integrating crop and livestock production, and employing agro-forestry technologies.

Post-harvest losses among resource-poor farmers are high, ranging from 20 percent to 40 percent, depending on such factors as the type of product, proximity of the farm to the homestead, and access to the market. Cutting post-harvest losses in economically efficient ways will augment the food supply and will also help increase farmers' incomes. Even at current production levels, research and education on nutrition can improve food use and safety, leading to a bigger impact on food security.

Programming priorities:

— Improving access, management, and administration of land.
— Diversifying and intensifying agricultural systems.
— Reducing post-harvest losses.
— Improving food safety, nutrition education, and use of available foods.

Agricultural sustainability and natural resource management

Degradation of natural resources (land, water, biological diversity) threatens the livelihoods of the poor, particularly in rural areas, where they rely on them more. Increasing poverty, in turn, limits the range of available options with regard to the sustainable management of these finite resources. The challenge is made more difficult by the increasing population densities in developing countries and the effects of climate change. Breakthroughs in agricultural productivity have been made in more favourable agro-ecologies where the use of inputs (irrigation, fertilizer, pesticides, and hybrid seeds) has been high. In addition to being inaccessible to most small farmers, however, agriculture based on high input use can easily produce ecological damage. Still, where intensification of agriculture based on modern technology was ignored, poverty and hunger have increased.

The sustainable development of agriculture will require a careful analysis and balancing of the potential tradeoffs between economic and environmental objectives. Programming in this area will be linked to Agency efforts under the Convention to Combat Desertification.

Programming priorities:

— Reversing current trends of land degradation.

— Promoting integrated natural resource management at farm, community, and watershed levels.

— Improving the efficiency and effectiveness of water use in agriculture.

Developing well-functioning markets

Food markets are responding to populations that are becoming increasingly urbanized, earning higher incomes, and demanding more diverse products of higher quality and value, including livestock products, fruits, vegetables, flowers, and processed foods. Also, as food markets globalize, trade in high-value agricultural products has been growing. Increasingly, food processors in developed and developing countries are sourcing their supplies globally. If the obstacles to international trade currently faced by the poorer developing countries-subsidies, and tariff and non-tariff trade barriers-can be overcome, agricultural producers have a greater chance to participate in such trade.

To help the rural poor participate in these local and international markets, agricultural output must diversify, quality of produce must improve, and agro-based processing must add value to primary products. Rural agricultural communities must also obtain greater access to credit. These communities can become better organized through cooperatives, which can help provide a range of necessary rural facilities, including those relating to input and output marketing, and financial services.

Programming priorities:

— Supporting agro-based processing and rural entrepreneurship.
— Strengthening local market organizations and institutions.
— Promoting agricultural services through cooperatives and rural agricultural education.
— Promoting access of farmers in developing countries to international markets.

Sequencing and targeting interventions

Development opportunities are not uniform among countries. Factors such as country income classification, extent of market liberalization, and trade policies usually influence agricultural development options. These parameters typically influence the type of assistance that may be attracted from international sources and indicate a country's policy and institutional readiness to pursue certain development options. Moreover, since all the identified priorities can rarely be undertaken at the same time, the Agency must also pursue interventions in an appropriate sequence.

The first priority is for the developing-country government to demonstrate commitment by creating the appropriate policy conditions. When that is in place, suitable technologies and adequate infrastructure work synergistically to help agricultural production expand. As these take hold, rapid benefits from improving asset distribution and providing agricultural services accrue to help increase the sector's productivity and competitiveness.

Agricultural programming options could be targeted along a continuum according to regional conditions within a country, from "safety net/self-sufficiency" approaches where conditions are poor, to market and export orientation where conditions are appropriate. Among the critical factors influencing conditions in a region are high versus low agricul\tural potential, good versus poor infrastructure, low wages and abundant labour versus high wages and scarce labour, and favourable versus unfavourable land distribution.

PROTECTING FARMLAND FOR AGRICULTURE

Sustainable agriculture is most often thought of as farming that is both economically and environmentally healthy, benefiting both food producers and consumers. There is a growing recognition that the protection of farmland around cities and towns - urban-influenced farmland-contributes to smart growth and the livability of every communities. Farms and farmland are valued as scenic landscapes and a part our heritage. They demand fewer public services and, therefore, cost taxpayers less than sprawling subdivisions.If protected as part of the "green infrastructure" around metropolitan areas, they can help guide suburban growth and promote urban revitalization.

It is equally important to recognize the interwoven relationship between smart growth and the sustainability of American agriculture. Not only does agricultural protection further smart growth, integral to smart growth is the protection of urban-influenced farmland. Sustainability begins--although it does not end—with the land that feeds us. The urban-influenced farmland contributes a significant amount of the U.S. food supply. Fifty-eight percent of the value of the food produced in this country comes from farms in counties within or adjacent to Metropolitan Statistical Areas, not from remote rural areas.Even more important, this includes over three-quarters of the fruits, vegetables and dairy products. A major reason is the high productivity and versatility of urban-influenced farmland.

Villages become sprawling cities, we squander this land at the risk of forcing agricultural production onto more fragile lands or overseas, diminishing the prospects of a sustainable U.S. agriculture. Urban-influenced farms are an economic bulwark against sprawl. This goes beyond the contribution that agricultural production makes to the local economy and the modest demand of farms for costly public services in comparison with the tax revenue they generate. Viable economic use of the open space around cities is necessary to justify effective land use regulation in a legal system that has become increasingly intolerant of "takings."

Because they cannot buy land around cities fast enough to influence development patterns over wide areas, sustaining agricultural use of that land is perhaps the best strategy that gives smart growth a fighting chance. Thus, supporting family farms and regional food systems becomes doubly important in urbaninfluenced areas. More people live near urbaninfluenced farms. This is stating the obvious, but there are several important implications that are easily overlooked.

One is that the environmental impact of farming this land arguably affects more Americans than that of any other agricultural land. Reducing farm runoff in the Corn Belt and restoring grasslands on the Great Plains are important environmental goals, particularly to those who live and farm there. But because more people are directly affected, the public benefits of promoting sustainable farming practices that result in clean water and abundant wildlife habitat are perhaps greater on urban-influenced farms than anywhere else in the country.

Agricultural practices still need improvement, but even now farm fields are almost always better for the environment than acres and acres of pavement.Once the land is paved, however, there is little opportunity to improve environmental quality. That is why New York City, for example, is helping upstate farmers protect the watershed from which the city draws its drinking water.But the deterioration of environmental quality can actually begin much earlier due to what has been called the "impermanence syndrome."

Farmers who are simply awaiting the developer's buy-out offer simply do not invest much in the upkeep or improvement of their operations, with results ranging from unsightly junk piles to increased pollution. Another implication of the fact that most people live near urban-influenced farms is that the rural landscape is highly accessible as an amenity that contributes to the quality-of-life.

People tend not to travel very far to take Sunday drives in the country, to visit pumpkin patches and Christmas tree farms, and to go hunting and fishing. And more and more of them are patronizing local and regional farmers' markets on a routine basis. The urban-suburban majority is glad the amber waves of grain and purple mountains majesty are still out there, but the countryside in their own backyard is where they spend more time. Thus, Americans care deeply about the loss of local farms to development. A recent national public opinion poll, for instance, found that setting aside open space around cities for farming was among the most popular smart growth strategies.Finally, to many Americans urban-influenced farms appear to be symbolic of the entire agriculture industry.

Modern production agriculture has more in common with large-scale manufacturing than with the Jeffersonian yeoman or Currier & Ives. Yet the public seems to be unaware of this, or at least to suspend disbelief, in expressing continued support for agriculture. Conceivably, it is because the landscape around most cities is still dotted with the small family farms many people want to associate with American agriculture. In substance and as a symbol, urbaninfluenced farms and farmland are far more important to the sustainability of American agriculture than is commonly acknowledged.

Without them, America would be a vastly different place. And if they disappear, American agriculture will have to undergo a radical adjustment, putting true sustainability farther out of reach. Indeed, trouble- some changes in American agriculture are already apparent, written on the landscape by the sprawl that is consuming urban-influenced farms faster than ever before. According to the latest U.S. Department of Agriculture statistics, over 2 million acres of rural land are being lost to development each year - double the rate of a decade ago.About 60 percent of this was agricultural land and most of the balance was in forests.

Moreover, per capita consumption of rural land appears to have increased by 50 percent in recent years,confirming that Americans are not just developing more farmland, they are wasting it on inefficient, low-density sprawl. This inefficient development of farmland has an impact that goes beyond the land actually consumed. For every acre paved over, another two or three acres can become riskier and more expensive to farm because of land use conflicts with new neighbors. Discontinuous sprawl makes the problem even worse by increasing the amount of "edge" between agriculture and residential areas. But several factors are particularly relevant because they help explain what must be done to protect Americans best farmland.

Foremost among these factors is that agriculture generally cannot compete with other enterprises in the marketplace for land. Corn chips simply aren't worth as much as silicon chips. But the competitive advantage of the development industry has been greatly exaggerated by government expenditures and policies that subsidize the construction of homes, shopping malls and factories. Highways, for example, inflate the price of the land along the right-of-way far beyond the ability of farmers to afford it, creating a bonanza for speculators while literally paving the way for sprawl.

US national policy of allowing homeowners to deduct mortgage interest from federal taxes—regardless of the size, value or location of houses—is a massive, direct subsidy to inefficient consumption of farmland that has become totally divorced from the professed policy objective of promoting affordable housing.In contrast to these powerful influences, local land use policies are woefully inadequate to control sprawl and, in most cases, simply reinforce the tendency of development to spread out over farmland.

The "A- 1" designation of most agricultural zoning ordinances in reality stands for "anything goes." Unless the playing field is leveled by changing these and other public policies, farmland will continue to be developed wastefully and indiscriminately. And the sustainability of American agriculture will continue to be compromised. On the other hand, their agriculture policies do little to help family farmers survive in urban-influenced areas. In recent years, about half of the income "earned" by farmers has come from the federal government in the form of subsidized loans, crop insurance discounts, disaster relief or outright payments under a program called "Freedom to Farm."But most of these government payments go to very large commodity producers, relatively few of whom farm in urban-influenced areas.

And while annual federal agriculture expenditures have increased five-fold since 1996 - from $6 to $30 billion - the share devoted to soil, water, wildlife and land conservation has declined from one-third to onetenth. The sum Congress has appropriated to help states and localities protect urban-influenced farmland - $40 million over 5 years - doesn't even amount to a rounding error in the federal agriculture budget. In short, the federal government does very little to serve the economic and conservation needs of farmers in urban-influenced areas, despite their significant contribution to agricultural production and the quality of life of metropolitan communities.

It is against this backdrop—low economic returns to farming and a policy framework that does little to help urban-influenced farmers, but much to hasten the transformation of farmland into sprawling subdivisions - that programs designed expressly to protect farmland are offering hope that agriculture can be sustained in urbaninfluenced areas. For the most part, these programs have originated with local communities, states and private conservation organizations like American Farmland Trust.

Best Practices for Farmland Protection

In the mid-1950s, Maryland became the first state to reduce property taxes on farmland to forestall its development. Every state has followed suit, reducing property taxes on farmland to a level commensurate with farm income. But many states have gone well beyond this necessary but insufficient step toward helping agriculture withstand sprawl. For example, California's landmark Williamson Act, passed in 1972, grants additional property tax relief to farmland owners who are willing to make a legal commitment not to develop their property for a decade or more.

Farmers who make such a commitment by enrolling in "agricultural security areas" in Pennsylvania and elsewhere are protected against nuisance lawsuits, special tax assessments and condemnation of their land for public purposes. These protections are important because they give farmers leverage over the construction of highways and other infrastructure that promote sprawl. Tax relief and agricultural security areas help stabilize urban-influenced agricultural use, but they neither prevent land speculation nor put the land off limits to development. Both these goals are achieved by conservation easements that permanently limit land development. These legal agreements are voluntarily entered into between landowners and either a government agency or private conservation organization. Farmers are compensated for giving up property rights. Payments average $1,500 per acre but can range upwards of $6,000.

Farmers typically invest the money in more land and farm improvements, use it to build a retirement nest egg or to assure the successful intergenerational transfer of farms.Thus, easements not only protect farmland, they also provide an infusion of capital to strengthen the agricultural economy. At last count, 19 states are buying conservation easements on farmland specifically to keep it in agricultural use. Quite a few local governments have followed suit, supplementing state funding with their own. Collectively, these programs-known as purchase of agricultural conservation easements ("PACE") or purchase of development rights ("PDR")-have permanently protected almost 900,000 acres of farmland by investing more than a billion dollars during the past two decades. Last year, PACE programs set records for both the amount of farmland protected (100,000 acres) and total funding ($160 million). But this isn't nearly enough to keep pace with the loss of farmland to sprawl. And it is a fraction of what is being spent by states and localities to protect land for open space, environmental and recreational purposes.

Private land trusts are also protecting farmland. The Land Trust Alliance reports that private organizations have protected an estimated 1.4 million acres of agricultural land with conservation easements, though it is not known how much of this is actively farmed or to what extent the purpose was to protect agriculture. Usually, land trusts do not pay cash for easements but instead convince landowners to donate their development rights in exchange for federal income and estate tax benefits that can enable landowners to recoup 50 percent or more of the value of the property rights they relinquish.

In contrast to easement purchases, however, donations appeal more to those who have off-farm assets and do not depend on agriculture for a living, than to full-time farmers who have few assets other than their land and cannot afford to give away what amounts to their retirement savings. Thus, those land trusts that have been most effective at protecting land for commercial agriculture have tapped into sources of public PACE funding, in effect, becoming acquisition agents for government. The value they add is their ability to act more quickly than public agencies, and their creativity in using easements along with other types of real estate transactions—for example, purchase-andleaseback, annuities and even limited development - to fashion solutions for individual farmers.

Indeed, the latest trend is for agricultural organizations to form their own land trusts. Conservation easements, agricultural security areas, tax relief and similar methods of protecting farmland are all voluntary and offer financial incentives to farmers in exchange for restrictions on land development. This explains why they are more popular in the farm community than the other basic approach to farmland protection - mandatory land use regulation - which many landowners consider an uncompensated "taking" of their private property rights.

But incentive programs have drawbacks, too: they are costly and slow to protect land. In their initial stages, they can do little more than produce a "checkerboard" pattern of protected and unprotected farmland. Thus, farmers still face the risk of land use conflicts with neighbors and uncertainty over whether enough farms will remain in an area to support businesses like farm equipment dealers that, in turn, support the farms.

Except in those rural areas where there is minimal development pressure, it is unrealistic to think that agriculture can be saved simply by purchasing development rights or using other incentives. When sprawl begins to threaten, the key issue becomes how to quickly stabilize agricultural land use over a wide area. One approach being taken by more local communities - and states - is to avoid planning and paying for roads, water and sewer systems in areas that they want to remain agricultural.

But communities are finding that even these additional incentives are not enough; that they must also resort to land use regulation to limit development of farmland. Some have opted for agricultural protection zoning that permits non-farm development only at a very low density so it will not conflict with commercial agricultural operations. The ability of local governments to adopt this kind of regulation depends on state enabling authority, which varies widely.

Attitudes change, however, and as sprawl causes more and more problems for cities and suburbs as well as rural areas, political and judicial support for effective local land use regulation seems to be on the rise. Though most mainstream agricultural institutions continue to resist regulation of any kind, many farmers appear to support regulation that protects their interests. For example, a recent American Farmland Trust survey showed that 58 percent of the nation's agricultural landowners would support restrictive zoning if it protects their "right to farm" against conflicts with encroaching development.

"Hybrid" Approaches

A handful of localities have overcome landowner resistance to effective agricultural protection zoning by linking it with the purchase of agricultural conservation easements as a way to compensate farmers and invest in the local farm economy. These "hybrid" programscombine incentives and regulations in such a way that the strengths of each counteract the weak- nesses of the other. Zoning regulations are quick and comprehensive, but they are temporary and, to many landowners, confiscatory.

However, effective regulations buy time for the community to raise money for easement purchases which, while slow and piecemeal, are also fairer to landowners and result in permanent protection of the land. The local communities that have taken this deliberate, balanced approach have been among the nation's most successful at protecting farmland and supporting a healthy agriculture in the face of sprawl. Not coincidentally, these communities also tend to be pioneers in what is now being called smart growth. Their objective, in most cases, was not just to protect a green, open landscape and the opportunity to farm, but also to help facilitate efficient, sustainable urban and suburban development.

Obviously, protecting farmland and sustaining agriculture around sprawling cities is not a simple task. There are many public and private institutions engaged in its pursuit, and even more ways in which funders could help them be more effective. So, it is risky to suggest only a few strategies as worthy of consideration. Nevertheless, it may be helpful to think of the agenda for the future in terms of four broad challenges. These frame the opportunity of funders to help farmers, communities, nonprofits and policymakers make a real difference by making strategic investments.

Farmland protection is a community affair. Unless people at the local level exercise their power to determine the fate of the land, everything else is just window dressing. The leading "hybrid" communities have shown how a balanced approach, using incentives and regulations, can produce results that are both effective and fair. But more communities need the motivation and skill to adapt this approach to their own circumstances. They need to understand the importance of farmland, both as an irreplaceable resource and as an ingredient of smart growth, so public awareness campaigns are critical. They need the practical tools to influence whether and how the land is developed.

Foundations need to support better understanding and awareness of effective planning and fair land use regulatory approaches. Another important role is to advances strategies that raise public funds for PACE programs that assure nobody bears an unfair burden. Above all, communities need the political will to take the steps that are truly necessary, not only to prevent sprawl, but also to protect agricultural land for the long run.

This kind of political will comes from a broad local consensus among those with most at stake.

The importance of PACE programs to farmland protection and smart growth cannot be overstated. Purchasing agricultural conservation easements is not simply a budget line item that competes with other open space priorities. It represents a commitment by society to share with farmers the cost of - and responsibility for - the protection and good stewardship of farmland, not only for food production, but also as scenic open space, unpaved watersheds and wildlife habitat.

Without such a commitment, it is understandable that farmers resist the kind of effective land use regulation that is necessary in many cases to forestall scattershot development of the countryside and to protect the environment. With such a commitment to shared responsibility, the property rights movement would be much less justified in pressing the claim that regulation creates hardship for farmers and ranchers. The doctrine of shared responsibility is a new way of looking at the property rights issue that has polarized the society and stalemated so many needed environmental and land use reforms, not least smart growth.

The leading "hybrid" farmland protection programs have all applied the doctrine of shared responsibility, combining incentives and regulation with synergistic results. So, in effect, have federal agricultural programs aimed at saving topsoil and protecting wetlands. The "sodbuster" provisions of the Food Security Act of 1985, for example, prohibit recipients of federal farm income support from plowing up highly-erodible land, while the Conservation Reserve Program provides incentives to farmers who agree to set aside such land from cultivation.

Chapter 5

Agricultural Sustainability

For agriculture, the issue of sustainability is linked to that of food security, i.e. the sustained ability of agriculture to provide adequate food supplies. Concern about food security stems from the fear that as population increases, our ability to meet increasing food needs will be limited by the natural resource base. In addition, the technology of the green revolution, which was the introduction from the late 1960s of high-yielding varieties of rice, wheat and maize, application of chemical fertilizers and modern pest control methods, coupled with increased capital investment in irrigation and on farms, may have exhausted its potential. Furthermore, second-generation problems, which are related to the high-technology package and agricultural intensification, are claimed to be undermining future productivity through soil, water, and genetic degradation.

Investment in irrigation infrastructure has also slowed down. Agriculture has encroached into wilderness lands, affecting biodiversity, which is fundamental to the sustainability of agriculture. The Food and Agriculture Organization of the United Nations (FAO) has estimated that between 1995 and 2010 the increase in agricultural cropland will place 85 million hectares (ha) of forests at risk. This trend of increasing threats to natural forests further exacerbates the possibility of climatic change through the release of additional carbon dioxide into the atmosphere. Indeed, the agricultural sector is being accused of undermining its own sustainability.

The issues concerning the possibility of sustained agricultural growth began in the 1940s and 1950s, when the physical availability of natural resources was thought to be a possible limit to future growth. In order to combat natural resource and environmental pressures, national and international research communities have joined forces in producing technologies that increase productivity, augment the existing natural resource endowment, and prevent food scarcity and starvation. The green-revolution technology was believed to be a win-win solution that overcame natural resource constraints and institutional changes.

The Brundtland Commission's definition of sustainable development, as cited above, applies to the concept in general. As far as agriculture is concerned, the Asian Development Bank (ADB) has another definition, namely "that which can evolve indefinitely toward greater productivity and human utility, enhance protection and conservation of the natural resource base, and ensure a favorable balance with the environment". This definition of sustainability is not just about maintaining environmental quality for a given level of resources. Nor is it about maintaining yields at current levels in perpetuity. The concept also includes (i) the need for enhancing productivity, and (ii) the need to meet increasing demands from growing populations. It is, therefore, not a static definition of constant production but refers to a sustained increase in production and consumption over time.

The ADB definition is particularly ambitious, considering that the standard economic interpretation of the Brundtland Commission's concept of sustainability requires that the per capita consumption of future generations remain at least as high as the current level. In order to maintain constant consumption levels over time, an amount equivalent to the economic depreciation of the exploited resources must be ploughed back into the investment as capital formation. This capital formation needed to replenish depleted stocks does not necessarily have to be physical capital. For agriculture, the ploughed-back amount could be in the form of investments in new technology and human resources. If consumption is allowed to increase over time, greater levels of plough-back investment are necessary.

IMPACT OF GREEN REVOLUTION

The green revolution has been central to Asia's agricultural success. A key element is the use of new "improved" crop varieties developed with the aid of modern plant breeding techniques. Before the Second World War, Japan and its then colonies were the only Asian economies to employ crossbreeding extensively to increase crop productivity. Similar efforts did not begin in the rest of Asia until 1950, at which time breeding programs were instituted almost simultaneously in most Asian countries. International breeding programs began shortly afterwards, for rice in 1960 at the International Rice Research Institute (IRRI), for maize and wheat in 1966 at the International Center for Maize and Wheat Improvement (CIMMYT), and for soybean, mungbean, and some major vegetables at the Asian Vegetable Research and Development Center (AVRDC).

Modern varieties (MVs), the new varieties developed through both national and international breeding programs, began to be released and diffused in Asia beyond Japan, the Republic of Korea, and Taipei,China, from 1965. The new varieties were generally superior in terms of yield potential, tolerance to pathogens and pests, and responsiveness to fertilizer and irrigation. They were also insensitive to photoperiod and/or required shorter growth time, making them more suitable for intensive cropping systems. The advent of the green revolution has saved Asia from famine and starvation. Nowhere has the impact of seed-fertilizer technology been greater than in South Asia where almost all countries have managed to feed their populations despite predictions of famine.

Bangladesh changed from being a net importer of 3.5 million metric tons (t) of grain annually in 1965 to selfsufficiency in grain by the early 1990s, by which time its population had grown from 53 million to 115 million. In India, where large-scale food shortages were avoided, the green revolution enabled food production to outpace population growth. Between 1970 and 1991, the annual rate of increase in food grain production was about 2.5 percent, while the annual rate of population growth over the same period was 2.2 percent.

Technology had enhanced the food-growing capacity of India to the extent that it could have fed an additional 350 million people during that same period. In the PRC, the food production index rose from 50 in 1975 to 145 in 1995, which implies that enough food was produced for an additional292 million people over that period. Similar success stories were repeated in Indonesia, Pakistan, Sri Lanka, and Viet Nam. Some countries, especially India, Thailand, and the Philippines, are now rapidly catching up with hybrid maize technology.

The green revolution not only helped to increase food production and supply, but also altered agricultural practices and cropping and trading patterns, and transformed rural livelihoods throughout Asia. The increased incomes and volume of trade encouraged associated activities such as food processing and transport. Expansion of electrically and mechanically powered irrigation and increased adoption of tractors and other farm machinery reduced the need for draft animals.

A villagelevel study in Punjab, India, covering 1965 to 1978, revealed that camels were no longer used as draft animals and that the use of bullocks had decreased substantially. In their place, the numbers of food animals such as buffaloes and goats increased. Milk and meat became more readily available for household consumption. Rural poverty in India declined substantially as a result of government spending related to the green revolution.

The impact of the green revolution on equity was questioned in early critiques. The technology involved can be seen to be selective and biased in favor of resource-rich regions and wealthy farmers. Fertilizer-responsive technology needs to be supported by a favorable environment, such as one with good irrigation, and tends to further aggravate the unequal distribution of income between resource-rich and resource-poor regions. Farmers also need credit worthiness, which tends to favor the large rather than the small farmer. Landless labor derives little benefit from these improvements, and employment levels have actually dropped due to the mechanization made possible by the higher productivity resulting from the green revolution. Rich and influential farmers were seen to maximize gains by ending tenancy agreements and lobbying for input and price subsidies.

The increase in the supply of labor-intensive crops kept real wages low, which helped to support the expansion of laborintensive enterprises. Fairbairn reviewed over 300 studies on the impact of the green revolution and found that 80 percent of the studies conducted between 1970 and 1989 concluded that the impact on equity of the green revolution was negative and that inequality increased during that period. It has been feared that the green revolution is a potential cause of increasing social antagonism and unrest.

The counter argument is that the negative effects on equity were the result of the early stages of the green revolution only. Citing field evidence from the northern Arcot region, Tamil Nadu, India, proponents of the green revolution indicated that the difference in yields between large and small farmers, evident in the 1970s, disappeared in the 1980s because smaller farmers were late adopters. In fact, small rice farmers and the landless made larger gains in family income than did large rice farmers, farmers of other crops, and nonagricultural households. There was no increase in the concentration of land ownership. One study that found widened regional disparities in India between the mid-1960s and 1970s also found that a second-generation effect of the green revolution was increased output and profitability of small farmers.

Other benefits have included widespread employment opportunities in postharvest operations such as storage, milling, marketing, and transportation. Increased rural incomes further brought about a diversification of rural economies and new opportunities for nonfarm activities. There was some loss in employment because of mechanization and the use of pumping for irrigation, but improvements in real wages led to increased earnings for the landless and nonagricultural households.

Another study on the impact of adoption of HYVs in seven Asian countries (Bangladesh, PRC, India, Indonesia, Nepal, the Philippines, and Thailand) concluded that although HYVs improved productivity in favorable (irrigated) areas relative to that in less favorable areas, other indirect effects have tended to prevent significant worsening of disparities in income distribution. These indirect effects have included increased real wages in unfavorable areas through migration out to favorable areas where employment opportunities are higher, decline in the real price of rice, which has benefited consumers, and changes in land tenure that have mitigated the worsening of disparities in income distribution.

An exception is the development of hybrid rice in the PRC where there has been a direct positive impact on equity, because the new rice was adopted in the mountains in unfavorable regions. A more recent study, on marginal returns to technology inputs in India in 1994, found that the marginal return in rainfed areas from government HYV expenditure was almost twice that in irrigated areas. On the basis of State-level data for 1970 to 1994, the authors confirmed that increased agricultural productivity reduced poverty directly by increasing farmer income and indirectly through employee wages and reduced agricultural prices. Poverty of the landless increased, although to a small extent.

Most studies on the impact of income distribution concentrate on income from rice farming. When the total income of all households, both those adopting and those not adopting the new technology, is considered, the impact of income distribution on rural households is negligible. Therefore, if rice is the only source of income considered, an inequality is to be expected because the nonadopters tend to reduce the amount of resources committed to rice production. By examining total income for all outputs, the impact on equity of the new technology is seen to be minimal.

Later critiques of the green revolution have focused on its ecological and biological impact. The high-technology package used has disturbed the ecological equilibrium, creating undue dependence on external inputs and stretching the Earth's support system beyond its capacity. The spread of HYVs, which have a narrow genetic base, increases the risk of greater exposure to pest and insect attacks. The associated intensive use of agrochemicals could have a negative impact on the quality of water, harming the health of farmers, consumers, farm animals, wildlife, and the environment.

Also, since the green-revolution technology concentrates on a few staple crops grown in favorable regions, farmers in unfavorable areas have no option but to engage in extensive agriculture, resulting in encroachment into natural forests and fragile ecosystems. The green-revolution package has inherent weaknesses and second-generation effects associated with its high input practices. Finally, the technology involved depends on fossil-fuel energy sources, which are nonrenewable. This could undermine the long-term sustainability of greenrevolution technology.

In 1985, The Consultative Group on International Agricultural Research (CGIAR) estimated that without the modern varieties about 20-40 million ha more would be needed to produce rice and maize in the humid tropics (CGIAR, 1985, cited in Harrington, 1993). The various criticisms should not be taken as a rejection of the green revolution or of the value of an increase in food supply and food security. Rather they should be taken as providing directions for future research and improvement.

Growth Trends in 1967-1997

The ability to meet increasing demands from growing populations requires that production growth exceeds population growth (Table 1). The green revolution has made this possible over the last few decades. Demand and supply projections up to 2010 indicate that production growth of cereals will be high enough to allow a slight fall in the real price of food. Rice is the only major staple crop for which prices may increase. Some other developments occurring in tandem with the green revolution have been innovations in the areas of livestock, aquaculture, and coastal and oceanic resources.

Asia contributes over 90 percent of the world's production of rice, about one third of all wheat and about one fifth of all coarse grain. Three major trends can be observed in the yields of field crops in Asia. First, growth in the production levels of food crops, mainly cereals and pulses, has been decelerating during the third decade of the green revolution (Table 1). The yield increases of cereals and pulses peaked at 3.8 percent per annum during 1977-1986, but slowed to 2.3 percent during the next decade. However, the latter growth rate was still above that of population growth for the decade. The average annual yield growth of crops other than cereals and pulses rose from almost zero from 1977 to 1986 to almost 2 percent in the following decade.

The second trend is a shift away from, or a fairly strong diversification out of, food grains in favor of higher value crops. The decline has been most drastic for rice, millet, and sorghum. The trend has also reduced the dominance of food grains in the total cropping system.

Table 1: Crop Production in Asia, 1977–1997

	Average Growth (percent per year)			
	Production		Yield	
	1977-1986	1987-1997	1977-1986	1987-1997
Cereals and pulses	3.82	2.60	3.80	2.29
Others	3.22	5.16	0.20	1.81
Total	3.47	4.08	2.51	2.71

[a] includes fibers, oils, roots, sugar, tea, coffee, tobacco, rubber, vegetables, fruits, and nuts.

Third, since the 1970s that the seed-fertilizer package is beneficial only in favorable environments, little progress on technology applicable to less favorable areas has been made. For example, rainfed rice yields are only half of those of irrigated agriculture, with even lower yields for upland and deepwater areas. Maize yields in Rajasthan, Uttar Pradesh, and Madya Pradesh in India and in many other countries are low. The exceptions are countries where rapid diffusion of hybrid maize has occurred, such as the PRC, Thailand, and the Philippines. Amongst the major cereals, wheat and maize continued to show robust yield growth during the last decade.

Rice is the only staple for which the yield growth, or the average annual increase in output per hectare, has fallen below 2 percent, almost half that of the preceding decade. The highest average annual rice yields in Asia were 6.564 and 6.545 t/ha in 1977 and 1997, respectively, both in the Republic of Korea. In the PRC, yields jumped to 6.187 t/ha from 3.704 t/ha. For Asia as a whole, yields went up from 2.596 to 3.840 t/ha during the same period. The Asiawide average, annual yield growth rate, however, did decline substantially, from 3.35 percent during 1977-1986 to 1.50 percent during 1977-1996. This deceleration reflects the fact that the growth potential of the early innovations of the green revolution has been exhausted in the best-suited areas.

IR8, developed by IRRI, was the first HYV to break the yield barrier for rice and started the green revolution outside the PRC. In fact, other HYVs have never been able to better the yields of IR8, although the progress made in pathogen and pest resistance in later HYVs has been remarkable. Considering that 74 percent of wet riceland is sown with MVs, there is now a renewed need for new rice varieties that will raise the yield potential.

Examining the rice industry more carefully also reveals that the slow down in yield growth could be related to the diversification of MVs into traditional varieties that have higher eating quality but lower yields. For example, yield declines in Karnal, Haryana, in the heart of India's green revolution riceland, were a result of the move to grow basmati rice. A similar but stronger trend has been observed along Asia's Pacific rim where a boom in manufacturing has increased the opportunity cost of agricultural labor. While the high-quality rice production area has expanded in response to greater demand and better prices, the production of low-quality rice and also of marginal food crops has declined.

However, migration from rice to more lucrative crops and to higher paying jobs in the city has also increased the cost of rice. Rapid economic growth in the cities has fuelled the demand for high-quality rice and horticultural products. In Thailand, where the industrial boom continued unabated for a decade before coming to an abrupt end in 1997, productive resources, especially labor, moved away from the agricultural sector in general and the rice sector in particular as a result of competition for these resources. Rapid growth in other newly industrialized economies has also placed pressure on rice production in terms of labor and land costs. Economic recessions will dampen demand somewhat, but this does not reduce the need for an effort to increase rice productivity.

In contrast with rice, the yield growth in wheat and maize remains strong. This is particularly true of wheat in the PRC,

India, and Pakistan. For maize, growth potential can be further tapped in favorable areas in Cambodia, India, Indonesia, the Philippines, and Viet Nam. The relatively strong growth trends in wheat are partly related to the success of breeding programs, which have raised the yield potential of new varieties at a rate of about 1 percent per year, while maintaining their resistance to the main pathogens. For example, the average number of newly released wheat varieties per year in India increased from 2.6 in the 1960s to 3.4 in 1970s and 7.2 during 1980 to 1985.

The yield potential of wheat MVs in South Asia has increased at a rate of 0.5 to 1 percent per year due to genetic improvement. The historical trend of productivity gains in the PRC has continued, with annual productivity gains averaging 3.33 percent from 1987 to 1997, a further suggestion that the sustainability of the yield potential for wheat should not be a great cause for concern. However, as with rice the profitability of wheat is decreasing, for example in areas like Karnal. This should be a greater cause for concern.

The rapid growth in maize production is a response to the boom in the poultry industry, which is a major consumer of maize. Elsewhere, hybrid maize was adopted much later, spearheaded mainly by the seed or feed businesses in the private sector. Recent successes in Southeast Asia, starting with open-pollinated varieties followed by hybrids, imply that the potential exists for an expansion of maize output and increase in yield potential in the tropical environment. Asia dominates the world in both rice production and rice consumption, accounting for over 90 percent of each. Rice is consumed universally in Asia, whereas wheat is consumed as a major part of the diet only in Pakistan and some parts of Bangladesh, PRC, India, Nepal, and Central Asia.

Wheat output in Asia is about 40 percent that of rice, and land sown with wheat covers about 60-70 million ha, about half the rice-growing area, 130 million ha. The amount of land sown with maize (35 million ha) is only about one third of that sown with rice. This relative order of magnitude is important when planning the allocation of R&D expenditure on food crops in Asia. Despite the fact that rice production in Asia is enormous, the internationally traded volume is small, accounting for only a small proportion of the total production-3 percent during 1994 to 1996-while 18 percent of wheat was traded internationally in the same period.

The small international rice market implies that a reduction in production by any major rice-consuming country will have a significant impact on international trade and that rice prices may potentially vary considerably. This in turn implies that the poor in Asia are relatively more vulnerable with respect to food prices. The low levels of foreign exchange held by some Asian governments following the financial and economic crisis in Asia may further aggravate food security. Sorghum, millet, and barley have shown a declining trend in production and area sown over the past 20 years, although yield growths have been positive and increasing at between 1.4 and 1.8 percent per year. Among other major crops, there has been an increase in oilseed production, particularly for soybean, rapeseed, sunflower, and castor during this period, while tubers have shown different growth trends.

Crop production is thought to affect the environment not only directly, through nutrient depletion and the emission of greenhouse gases, but also indirectly where expansion of crop areas is a threat to forest areas.

Perennial crops

The major perennial crops in Asia, as measured in terms of harvested area (above 1 million ha) in 1997, are coconuts (9.1 million ha), sugar cane (8.6 million ha), rubber (6.7 million ha), oil palm (4.7 million ha), tea (2 million ha), and coffee (1.6 million ha). Together, the harvested area of coconuts and sugar cane (17.7 million ha) is less than half of the area devoted to maize and close in size to the area devoted to fruits (16.8 million ha) and vegetables (20.8 million ha). In Asia, perennial crops may be grown on large plantations, in smallholdings, or under a subcontracting system as is the case with the Thai sugar cane industry.

Tree-based systems, when properly managed either as plantations or agroforestry systems, have a relatively benign impact on the environment, especially when grown in the uplands and on steeper slopes. Also, once an investment is made in trees, the land is committed to a fixed pattern and therefore cannot be easily converted into land for food crops, at least in the short term.

The largest producers of coconuts in 1997 were Indonesia (14.7 million t), the Philippines (12.1 million t), India (9.8 million t), and Sri Lanka (2 million t). In the last decade, a small decline in harvest areas has occurred only in the Philippines. The World Bank has forecast a substantial decline in prices for both copra and coconut oil until 2010. Southeast Asia produces almost all of the world's rubber output of 5.2 million t. Thailand is the largest producer, contributing on average about 2 million t of output, which is about one third of the world's total natural rubber production, from harvested areas that covered 1.5 million ha in 1997.

The second largest producer, Indonesia, produces on average 1.5 million t per year from 2.3 million ha of plantations, followed by Malaysia, with about 1 million t from 1.5 million ha. These three countries together produce about three quarters of the world's total output. Harvested areas in Malaysia are experiencing a declining output, while Indonesia has seen a slight increase over the last seven years. Countries showing strong increasing trends but from relatively small bases are Viet Nam, Myanmar, and Cambodia. The World Bank's forecast for rubber prices indicates a shortterm price drop followed by a recovery towards 2010.

Tea is a traditional crop in many Asian countries, but it is grown on a relatively large scale only in the PRC (0.9 million ha), India (0.4 million ha), Sri Lanka (0.19 million ha), and Indonesia (0.11 million ha). The total area used for tea production has been relatively stagnant since 1975. India and Indonesia are the only countries showing slight increases in harvest areas. Sri Lanka, a worldrenowned producer of black tea, has experienced a clearly decreasing trend in area under tea. Harvested areas for coffee in Asia total 1.6 million ha and almost all of this is in Southeast Asia, i.e. Indonesia (0.8 million ha), India (0.24 million ha), Viet Nam (0.19 million ha), and the Philippines (0.15 million ha). The area under coffee is increasing strongly in Indonesia and Viet Nam and slightly in India.

Oil palm is not yet a very widespread crop in Asia, but the area under harvest is growing rapidly in Indonesia, Thailand, and Malaysia. The crop has high yield potential and requires a relatively small amount of labor for planting, maintenance, and harvesting. It demands a warm climate and evenly distributed rainfall, making Indonesia, Malaysia, and southern Thailand suitable growing areas. In 1997, the industry suffered from the widespread fires in Sumatra and Kalimantan, Indonesia. It can be expected that further expansion of oil palm will be somewhat hindered by the financial and economic crises occurring in the three major producing countries. Moreover, the World Bank has forecast a rapid decrease in the price of palm oil until the year 2010 due to excess competition and production.

Land clearing for agricultural tree plantations in Sumatra was proven to be partly responsible for the fires in 1997 that erupted into a regional environmental problem. The fires, exacerbated by the long drought associated with El-Niño, lasted from mid-1997 to early 1998, producing enormous quantitiesof smoke and haze that covered Indonesia, Singapore, Malaysia, and southern Thailand. In Indonesia, 5 million ha of forest and agricultural lands were damaged and 70 million people in the region were affected. The haze also seriously affected the previously booming tourism industry in the four countries. Total damages to Indonesia were estimated at $3.8 billion1 while about $670 million worth of damage was done to neighboring countries. Development policies favoring the conversion of forest into plantation have been seen as a major contributor to the haze problem.

Well-managed tree plantations tend to be less harmful to the environment than some field and garden crops. Wellmanaged plantations with good ground cover can reduce the rate of run-off and erosion to below 5 t/ha per year, which is a better rate than that of degraded forests and shrubs. For example, a well-managed tea plantation results in an annual loss of only 0.24 t/ha of soil, compared with 25-100 t/ha for vegetables, potatoes, and tobacco, and 0.3 t/ha for dense forest. In the early stages of establishment, erosion rates tend to be high but the mulching done during the first two years of planting can considerably reduce run-off and soil erosion.

Asia is a big producer and consumer of sugar. It has five of the world's top ten consumers, namely India, PRC, Indonesia, Pakistan, and Japan. The top five producers of sugar in 1997 were India (277 million t), PRC (82.57 million t), Thailand (45.85 million t), Pakistan (42 million t), and Indonesia (27.76 million t). The Philippines was a much bigger producer than both Pakistan and Thailand in the 1970s and early 1980s but the latter two countries increased their capacity substantially during the 1990s. Asia produces about one third and consumes about 45 percent of the world's sugar output. Hence, the region has a sugar deficit. The annual rate of growth of sugar cane production has also declined, from 4.9 percent in the 1950s to 1.6 percent in the 1990s.

The sugar cane industry in Asia is dominated by smallholders, implying a need for an efficient institutional arrangement between them and factories. Yields are highest for Indonesia and the PRC (about 71 and 75 t/ha, respectively), both of which benefit from irrigation. Yields in the Philippines and Thailand, which use relatively low-input rainfed systems, are about 69 and 49 t/ha, respectively. However, Thailand has the highest average sucrose content at 13 percent, followed by India (12 percent), Philippines (11 percent), and Indonesia (9.6 percent). Thailand has the lowest production cost, followed by India and Indonesia.

Forest plantations

Forest plantations in the region, such as teak plantations in India, Myanmar, and Thailand, were established early in the 20th century. These plantations were established mainly by government organizations for several reasons, including production and conservation. The harvesting rotations vary from medium, i.e., 20-30 years, to long, 40-80 years, aiming primarily to produce high-quality timber for the international market. Currently, most of the plantations are State owned.

Since 1980, the private sector has become increasingly involved in setting up large-scale plantations of species that are especially fast growing, such as Acacia, Albizzia, Eucalyptus, Gmelina, Paraserianthes, and bamboo in the PRC, India, Indonesia, Malaysia (Sabah and Sarawak), the Philippines, and Thailand. The primary objective of these commercial or private plantations is to produce industrial raw material, including roundwood and pulpwood, with short and medium harvesting rotations of 10-20 years. This objective is being promoted by the governments concerned. With very few exceptions, modern breeding techniques and improved clones are increasingly used in Asian plantations. There are also a few examples of manmade forests, established by indigenous people using traditional knowledge, and which have gained national and international recognition.

Forest plantations in Asia in 1995 totaled 59 million ha or about 15 percent of the area of natural forests. Between 1980 and 1995, Asia lost 63.26 million ha of natural forest cover (FAO, 1995b, 1997a). The total area devoted to forest plantations is about nine times that occupied by rubber plantations. Despite the fact that about four fifths of the existing plantation areasare used for nonindustrial purposes, e.g. for conservation and for household consumption or community uses, forest plantation establishment rates have lagged far behind deforestation rates.

The PRC has the largest plantation area, with an annual planting rate of 1.1 million ha. Most of the PRC's plantations are aimed at conservation and nonindustrial purposes. Recent devastating flooding has prompted the Government of the PRC to stop commercial logging in the western watersheds. India has the largest area of industrial plantations, totaling 5.7 million ha, consisting of fast-growing species (5-10 year rotation) covering 0.9 million ha, and other industrial species covering 4.8 million ha. In Southeast Asia, Indonesia has the largest area of forest plantations, 6.1 million ha, one third of which is planted for industrial purposes. Lao PDR lies at theother extreme, having the smallest plantation area, 4,000 ha, most of which is recent and geared towards the pulp industry. Among developing countries, the proportion of industrial to nonindustrial plantation area is highest in Lao PDR, Myanmar, Malaysia, and Sri Lanka.

Although there are many different tree species planted, especially in tropical Asia and the Pacific countries, eucalyptus appears to be the preferred species group of regional planting programs. This species group accounts for about 16 percent (5.2 million ha) of the total plantation area in the region. Acacia, teak, and pine are also major groups, accounting for 11 percent (3.4 million ha), 6 percent (2.19 million ha), and 4 percent (1.25 million ha) of total plantation area, respectively (FAO, 1995b). The remaining 64 percent (20.6 million ha) of plantation area contains other or unclassified tree species such as Albizzia, Dalbergia, Casuarina, Leucaena, Swietenia, Xylia, Gmelina, and Pterocarpus.

Eucalyptus and acacia plantations are established primarily for the pulpwood and medium-fiber wood industries. They are also favorites of nonindustrial plantations established for community use, e.g. for fuelwood, small wooden poles, for land rehabilitation, and for environmental conservation. The eucalyptus species planted in the region include Eucalyptus camaldulensis, E. deglupta, E. europhylla, E. globulus, and E. grandis, the hybrid species. The most common acacias are Acacia auriculiformis, A. mangium, and A. nilotica, and the pines most often found are Pinus kesiya, P. merkusii, P. caribaea, and P. oocarpa.

The growth and yields of these species vary according to onsite characteristics and genetic material (improved seed sources and clones). In Indian and Indonesian teak plantations, the yield of an average plantation is about 2-3 m3/ha/year after about 50-70 years. The average yield of eucalyptus plantations is about 6 m3/ha/year after 8-10 years. The average yield of Acacia mangium plantations in Malaysia and Indonesia is more than 20 m3/ha/year after 5 years.

Scientific progress relating to the establishment of forest plantations and logging operations has been stagnant because government agencies or State enterprises run these activities. Unlike forest plantations in developed countries or those operated by multinational corporations, forest plantations in Asia are run under low-input, low-output systems. Exceptions are found in the large-scale plantations in Sabah.

The impact of forest plantations on soil erosion is indicated in Table 2. Surface erosion from well-managed forest plantations is small, averaging around 0.6 t/ha/year. However, if forest litter is removed for use as fuel, as is the case in the PRC, erosion can be much greater than for shifting cultivation.The impact of forest plantations also depends on the species planted. Villagers often find that Eucalyptus camaldulensis plantations tend to lower the water table and dry up shallow wells, as these forests can absorb up to ten times more water than pioneer forests and four times more water than secondary forests.

Table 2: Surface Erosion in Tropical Forests and Tree Plantations

Forest and Tree Plantations	Annual Soil Loss (t/ha) Range	Mean
1. Natural forests	0.03–6.20	0.3
2. Shifting cultivation during fallow-period years	0.05–7.40	0.2
3. Forest plantations	0.02–6.20	0.6
4. Multistoried tree gardens	0.01–0.15	0.1
5. Tree plantations with cover crop/mulch	0.10–5.60	0.8
6. Shifting cultivation cropping	0.40–70.00	2.8
7. Agricultural intercropping in young forest plantations	0.60–17.40	5.2
8. Tree plantations, clean weeded	1.20–183.00	48.0
9. Forest plantations, litter removed or burned	5.90–105.00	53.0

National and international attention has increasingly been paid to the possibility of combining tree species with field crops, i.e. agroforestry, for small landholders. The International Center for Research in Agroforestry (ICRAF), which is based in Kenya, has expanded its research activities to cover Asia. In Indonesia, about 70 percent of the total rubber is produced under agroforestry systems. These systems have been acknowledged as providing sustainable support in areas where the soils are too poor to grow food crops on a continuous basis. Agroforestry is estimated to have been adopted by about 7 million people and occupies approximately 2.5 million ha of land in Sumatra and Kalimantan, and includes the damar agroforest.

Today, there are diverse and complex agroforestry systems that mix perennials with food crops. Some of the agroforests provide ecological functions similar to those of natural secondary forests, such as carbon sinks, sources of biodiversity, and means of alleviating soil erosion and flooding peaks. The development and adoption of agroforestry is largely dependent on various physical, environmental, political, social, and economic conditions. High population pressure combined with low per capita income and forest resources that are inadequate for local needs (including both timber and nontimber products) has necessitated the use of fuelwood and provides increased incentives for agroforestry.

An abundance of State land under a strong land-use policy, together with government incentives and support, reduces the cost of and increases returns from agroforestry. However, the small amount of arable agricultural land per capita may limit the profitability of treebased systems. Agroforestry has been adopted and practiced widely in the PRC, India, Indonesia, Lao PDR, and Viet Nam. With the exception of Lao PDR, the population density in these countries is high, the income per capita low, and forest resources scarce. In India, for example, the annual requirement for fuelwood and timber is 220 million t and 280 million m3, respectively, while the sustainable production levels of these products from the forests is only 30 million t and 12 million m3, respectively. To meet the demand for such forestry products as well as to maintain sustained production from the forestry and agricultural sectors, agroforestry has been promoted in various parts of India.

The role of agroforestry in fuelwood supply is very important in both rural and urban areas of Bangladesh, where 90 percent of the fuelwood used is from home gardens. Agroforestry development in Lao PDR has been a response to government policy on land taxes and the desire of individuals to claim more land, especially land that is easily accessible. In the PRC, agroforestry has the support of both central and local governments, with the aim of improving environmental and economic conditions. The systems used in the PRC are much more diverse than those in any other country in the region and include home gardens, strip shelter forests in combination with cropping systems, woodlot plantations, and tree shading with cropping systems.

To address the shortage of fuelwood, more than 3 million ha of woodlot plantations have been established with an annual production of 10-30 t/ha, depending on the species planted. The shelter-forest system also increases wheat yields and latex production in rubber plantations. Further, it generates over 6 million m3 of timber and 3 million m3 of fuelwood per year through thinning and final harvesting.

The system in Lao PDR has been initiated mostly under the taungya system which involves planting teak or paper mulberry in combination with rice, pineapple, maize, or other cash crops to improve shifting cultivation. This is especially so in the northern part of the country where large shiftingcultivation areas are being replaced by teak. Although agroforestry has been adopted and practiced successfully in many countries, the scale of its practice is limited to subsistence production. There are many factors that limit its growth and development, including insufficient mechanisms for the exchange of agroforestry experiences, inadequate policies, rules, and regulations, and an insufficiency or a lack of incentives from governments. Other factors include a lack of access to state-of-the-art technical information and farmer-generated knowledge among small households, a lack of up-to-date market information on agroforestry products, and inadequate support services for expanding agroforestry activities.

Livestock production

Historically, livestock were raised using resources that were of little value for other uses, such as household food wastes or land that was not fertile enough for crops. Today, livestock production has emerged as one of the more advanced segments of agriculture, and also as one of the most important components of global agriculture. Of the three major production systems (land-based grazing, mixed farming, and industrial farming), industrial farming has seen the most rapid changes in the last decade, with most of the growth occurring in developing countries.

In Asia, where industrial and urban growth has been particularly rapid in the last decade, there has been a remarkable expansion of the peri-urban poultry and pig industries. This growth has been a response to increased demand as well as to technical advances. More concentrated feed and improved animal health have resulted in a more favorable feed-conversion ratio, giving these commodities a higher return on investment. The structural shift also reflects the pressure on land from population increase, rendering the expansion of land-intensive animal husbandry more difficult.

Livestock production is vital to the overall development goals of Asian countries: livestock production improves food security and nutrition and increases employment. Modern industrial farming operates side by side with small-scale farming, although there is still a predominance of mixed farming systems. Some 90 percent of livestock production in the developing countries of Asia comes from smallholders or landless persons.

Traditional farming is based on systems with minimal or no imported inputs and where livestock and crop activities are integrated. Farm products are mainly for domestic consumption and the excess is sold locally. The demand for livestock products has increased rapidly in urban centers, and traditional landbased livestock production with limited use of resources is unable to meet this increasing demand. The commercial nature of livestock production encourages specialized intensive farms to move nearer to the market place in the city, i.e. in peri-urban areas.

Smallholder farms, because of their noncommercial nature, have little access to credit facilities or modern technologies to enhance their activities, further losing their competitiveness, resulting in their being supplanted and marginalized. In some countries, rural producers, because of a lack of infrastructure, economies of scale, and insufficient marketing facilities, face heavy competition from urban producers, which often limits rural livestock production to subsistence levels.

In rural areas, there are many animals with a low level of productivity because they are only being fed at about maintenance level. With more feed, much of the additional nutrition would go directly to production. This has been demonstrated in Bangladesh where the provision of a small amount of supplemental nutrition has led to an increase in milk production from 1 to 6 liters per day in indigenous cows. Increasing livestock production for poor farmers would provide a useful short- to medium-term benefit, especially where farm labor is underutilized. If animals already owned by the farmer become more productive, the benefit is received at little additional cost. Other constraints to livestock development in Asia for smallholders include the scarcity of feeds, high incidence of animal and poultry diseases, the prevalence of traditional livestock management systems, and inadequate access to credit.

Low livestock productivity can also be the result of poor animal husbandry. Appropriate livestock practices can make a contribution to raising productivity, but farmers often have limited access to education and training. Low productivity can also sometimes be attributed to the fact that these small-scale farmers do not have access to the better breeds of livestock. However, replacing local cattle with improved breeds will not solve the productivity problem unless feed resources on the farms are increased.

In the past, it has been government policy in many countries in Asia to focus on importing foreign breeds in order to achieve higher productivity levels. However, establishing a modern industry with imported breeds has proven difficult. This has been largely due to the fact that the greatest improvements that can be derived in animal productivity require better management and improvement of feed resources, not simply a different breed. Foreign breeds will most likely have different feed requirements that cannot be met by smallscale farmers and may not adapt to local conditions as well as do the indigenous breeds. Additionally, if feed requirements are not met for the foreign breeds, their reproduction rate may be affected.

Asia has relatively few major infectious disease problems in cattle and buffalo. However, foot-and-mouth disease has been difficult to prevent amongst countries with easily accessible land borders, unless joint programs have been implemented. In some Asian countries, notably Bangladesh, Indonesia, Philippines, and Thailand, the poultry sector is very important and the control of disease, especially Newcastle disease, is crucial, particularly at the village level. In Bangladesh, the estimated annual loss from poultry diseases is $240 million, of which about half is attributable to Newcastle disease alone. The increase in industrial livestock production has also increased the rate of livestock disease; diseases also spread more quickly and are harder to contain in intensive animal production.

In most developing countries, priority in the past has been given to the production of food crops. Several considerations highlight the current need to give greater priority to livestock development. While in the industrial world the demand for milk and meat will likely plateau, or even decline, in the developing world population growth and urbanization will fuel a strong increase in demand. For example, between 1986 and 1996, growth rates for meat in Asia were 7.7 percent, compared with only 0.7 percent for the rest of the world. Milk growth rates over the same period in Asia were 4 percent, whereas annual growth rates were negative at -0.8 percent for the rest of the world. Current levels of milk and meat consumption in developing countries are only about one fifth of those in developed countries.For the region as a whole, a strong trend has been observed towards the incorporation of more and more animal protein into the population's traditional vegetable-based diet.

Connected with this trend is an increasing selectivity as to which parts of an animal are used for food. Traditionally, most parts of an animal were utilized, even if there was much wastage due to insufficient recovery and re-utilization technologies. Now the global trend is clearly to meat, and more often to lean meat. Other products, such as offal, blood, and bones, are increasingly used industrially and often recycled as feed.

The demand for livestock products is also highly incomeelastic, and thus demand will increase with rising incomes. In Asia, the demand for meat is expected to triple by 2020. An accelerated livestock production program to satisfy this growing demand for primary livestock products such as milk, meat, and eggs will be required. Growth of all livestock sectors has been high over the past two decades, and this has been especially true of poultry and eggs. It will be a challenge for the livestock sector to satisfy future demands while at the same time preserving the natural resource base. The long-term objective is to produce and supply sufficient and safe animal protein for rapidly growing and urbanizing populations, under socially and environmentally acceptable terms.

Growth in meat production until 2010 is expected to come from increases in productivity and from greater numbers of animals, with these factors accounting for about one third and two thirds of growth, respectively. Poultry production is projected to rise most rapidly, followed by pork production. More than 90 percent of the increase in pork production, however, will come from one region, East Asia, including the PRC. There are some differences between the major regions in structural changes, although the main trends are common to all. The proportion of poultry meat to total meat output is expected to continue to increase in every region while the contribution of cattle and buffalo meat will likely decline. Yields per animal are also expected to grow faster than in the past twenty years as a consequence of improvements in health, feed, and pasture carrying capacity.

Although the contribution of the livestock sector to GDP is relatively small, it is nevertheless significant in terms of the total output of the agricultural sector. The sector in Asia contributes from about 10 percent to about one third of agriculture's gross added value. However, livestock statistics generally quantify the products that are eaten and traded such as meat, milk, and eggs, and do not consider products such as draft power and manure (as fertilizer, fuel, or feed). As a result, the statistics greatly underestimate the role and importance of livestock.

Growth rates have increased for all major livestock products over the past two decades. Productivity increases have been the main source of production increases, as opposed to the expansion of the industry or an increase in livestock numbers. Growth trends also differ within Asia, as there are subregional variations in production and consumption across the region. For example, South Asia consumes large quantities of milk but little meat. Farming is also still largely small scale with manure and draft power remaining highly important. East and Southeast Asia have seen the greatest increase in intensive production of monogastric animals such as pigs and poultry. In Thailand, poultry has become a major export earner.

The recent financial and economic crisis in Asia has also affected the industry, particularly in countries that rely on imported feed. These countries have seen a sharp increase in the prices of imported feed and other inputs as well as a contraction of demand. This has caused a drop in industrial production and in some cases it has been wiped out altogether. Industrial livestock production had grown rapidly in Indonesia until the crisis: large-scale poultry production, which was growing very fast there, was heavily damaged by the crisis, mainly because it is heavily dependent on imported feed. In

Malaysia, although livestock production growth rates remain high, the industry also depends largely on imported feed.

Due to a lack of livestock feed resources and shortage of land for livestock feed production, most countries in the region are not in a position to develop large-scale intensive livestock industries without relying on imported feed. However, some countries have turned to domestic feed resources. Thailand, for example, has now increased its use of cassava for livestock feed. Such developments offer a window of opportunity for the development of commercial smallholder livestock because the competitive pressure from large-scale industrial producers has, to a large extent, subsided.

In terms of total meat, less than 3 percent is produced under the grazing system. The bulk of meat production originates from mixed systems that account for two thirds of total production. Among the mixed farming systems, rainfed systems only account for one seventh, while the emerging industrial system now accounts for one quarter of total meat production. About 21 percent of the world's arable land is producing feed for the livestock industry, and livestock production uses 32 percent of total cereal production. Maize accounts for half of all feed grain, with barley and wheat as the other main components. Soybean is the most important component among oilmeals.

Milk production shows a similar trend to that of meat production, although at lower rates. The highest per capita production is in South Asia, where India and Pakistan are leading, although growth in the industry is now slowing there. The PRC and Thailand have also shown strong growth rates, but per capita production is still low. Strong growth has continued for poultry and egg production, averaging 7 percent or more per annum during the last 30 years. Malaysia and Thailand have had strong growth in both areas, and although the PRC has also shown strong growth rates in both areas, production per capita of poultry is still low. Japan has high egg production per capita and moderately high poultry production per capita. but growth rates are low because local demand is satisfied by imports, and the industry is now stabilizing.

Areas of South Asia have experienced strong growth in livestock production over the past 30 years. Strong growth in poultry has been manifested in India, Bangladesh, and Pakistan. In India, livestock production is a high priority at every administrative level for two primary reasons: production self-sufficiency and rural development. Small farm size remains the biggest obstacle to development of the livestock industry there.

In Bangladesh there is a large and important livestock population, but the animals have low productivity, mainly due to the insufficient quality of feed resources and a high incidence of disease. Bangladesh is deficient in livestock production but cannot afford many imports. However, growth of the industry has been improving over the past 30 years.

Production has remained low in other countries in South Asia such as Nepal, where 90 percent of the population depends on integrated farming, including livestock raising. Feed resources need to be developed, which would lead to more productive animals. Difficult transportation conditions and weak marketing facilities are among other constraints facing Nepal today. However, egg and poultry production have shown moderate increases in growth rates since 1975.

Central Asia has the highest per capita production of milk and a high production of meat per capita. However, the production of other livestock products is low. Data have not been available to determine growth rate trends over the past three decades for Central Asia. Livestock production has a great potential for contributing to either the degradation or the enhancement of the environment, depending on the technologies and practices adopted. The environmental challenges range from overgrazing (degradation and erosion of land), to deforestation, to degradation and pollution of water resources, emission of greenhouse gases, and loss of biodiversity.

Growth rates in the more developed countries such as Hong Kong, China; Japan; and Singapore have begun to level off. Land scarcity in Hong Kong, China; and Singapore dictates that they will remain largely importers of livestock products. Recently, the governments in these countries have paid more attention to environmental pollution and health risks resulting from livestock farming than to increasing production capabilities. The threat to human health from livestock production is exemplified by the spreading of a rare virus from pigs to humans in Malaysia in early 1999. More than 100 people died and almost one million pigs were destroyed to keep the virus from spreading.

Recent rapid growth of the livestock sector near urban areas has created additional pollution problems for already congested and relatively highly polluted areas. Wastes generated by intensive production units, especially pig production, are a major source of water, land, and air (odor) pollution. Heavy metals arising from feed can affect the health of nearby residents if wastes are not treated properly. The use of additional water to clean solid waste from production units further increases the amount of waste to be treated and in turn the cost of wastewater treatment.

Animal waste from intensive systems can be used to produce biogas for heating, drying, and power generation. The digester effluent can also be used to fertilize crops and fertilize/feed fishponds. Appropriate government intervention in this area would allow environmental problems to be solved and maintain long-run production, while deriving benefits from biogas energy and nutrient recycling. Other initiatives, such as limitations on stocking density, would help to reduce pollution discharges.

Pressure for greater emphasis on environmental conservation is evident throughout the region. The greatest threat is from overgrazing and in many countries, especially in semi-arid areas, livestock numbers already exceed the carrying capacity of unimproved natural grasslands. Overgrazing is a major concern, especially in India and Central and inner Asia.

It will be important to identify situations where the raising of livestock is out of balance with the adsorptive capacity of the soil, water, and air. Competition between crops and grazing generally results in resource degradation and finally in the collapse of livestock production, especially for the larger ruminants. In some areas, there is a need to restrict the density of animals raised, the objective being to optimize the long-term productivity of the land as a whole while maintaining ecological diversity and environmental balance.

In tropical Asia, the problem of deforestation is complicated as it is intertwined not only with livestock production, but also with logging operations and human population pressure, especially in areas where suitable land is scarce. Overgrazing, which leads to land degradation, also leads to deforestation as new land is cleared for use. Of the livestock production systems, mixed farming tends be the most environmentally friendly: it allows reuse of animal waste as organic fertilizers that replenish the soil and reduce erosion. Mixed farming also provides protection against product/price fluctuation risks. Each year, livestock produce about 13 billion t of waste, and supply 22 percent of total nitrogen fertilizer and 38 percent of phosphates of animal origin.

Crop waste can also be reused as animal feed. In livestock systems other than mixed farming, transport costs from nutrient-surplus to nutrient-deficient areas can be high. For example, with intensive systems, there is a need for transport of feed to and manure from production areas. It may eventually be less costly to relocate intensive production areas away from peri-urban centers, if a proper transportation infrastructure can be developed. Some rapidly expanding urban centers such as Ho Chi Minh City, Viet Nam, have already initiated programs to move intensive pig and poultry production outside city boundaries. The difficulties in transition from extensive to more intensive livestock production have also led to increased environmental degradation.

There have been massive investments in past decades in physical infrastructure, including the construction of roads, ports, and communication facilities, but also of slaughterhouses, and cold storage and retail facilities. All of these have greatly reduced transaction costs and have, in numerous places, made possible trade in livestock products. Because of the perishable nature of livestock products, infrastructure development has an extremely stimulating effect on livestock production.

As per capita incomes increase, more animal products pass through market and processing channels before consumption. This leads to even greater waste production. The most important environmental impact of animal product processing results from the discharge of wastewater. Additionally, heavy metals such as copper, zinc, and cadmium are used as growth stimulants in some feeds. Without proper management of these discharges, intensive farming systems discharge waste containing heavy metals at levels that are harmful to animals and human health. At present, only the Organisation for Economic Co-operation and Development has regulations that aim to reduce the level of heavy metals in feeds.

Fishery production

Asia accounted for more than half of total world fishery production in 1996. During the two decades ending in 1996, total production of fish and shellfish in Asia increased at a much faster pace (approximately 4 to 5 percent per year) than did the production of food crops, rising from nearly 28 million t to 67 million t from 1975 to 1996, raising Asia's share of the world total from 42 percent to 55 percent (Table 3). The overall performance of Asian fisheries is especially remarkable when their growth is compared with that of the rest of the world. Asian fishery production grew at an average 4.8 percent per year over the past decade, up from 3.6 percent a decade earlier, whereas that of the rest of the world declined from 2.6 to 0.3 percent over the same period (Table 3). Asia was the main force propelling the overall increasing trend in world fishery production.

Table 3: Asia and the World: Fish and Shellfish Production, Selected Years

	Output (t, million)						Growth (%)	
	1950	1975	1985	1990	1995	1996	1977–1986	1987–1996
Asia	6.46	28.15	38.47	47.45	63.28	67.11	3.6	4.8
Rest of the World	12.74	37.59	48.67	51.56	54.00	53.90	2.6	0.3
World Total	19.20	65.74	87.14	99.01	117.28	121.01	3.0	2.6
Asia's Share in world (%)	33.7	42.8	44.1	47.9	54.0	55.5		

Two striking features of Asia's fisheries growth during the past decade are the emergence of the PRC as the predominant producer, and an increasing contribution from aquaculture. The PRC contributed nearly half of Asia's total fish and shellfish production in 1996, compared with only 16 and 21 percent in 1976 and 1986, respectively. However, only a few have registered annual growth rates above 3 percent, while production in Japan and the Republic of Korea has declined. Japan, Asia's largest fish producer until 1988, now takes second place after the PRC. Japan's share in Asia's total fish and shellfish production dropped from 34 percent in 1976 to merely 10 percent in 1996.

Much of the growth of Asian fisheries during the past decade was fuelled by aquaculture production, which grew by over 11 percent per year between 1987 and 1996 in both freshwater and marine waters. Once again, most of this growth has taken place in the PRC, where fish production increased by a factor of 6.6 during 1977 to 1996, with particularly high growth, an annual average of 13 percent, during the past decade. Over half of the PRC's fish and shellfish production in 1996 came from aquaculture, and accounted for 75 percent of Asia's aquaculture production of fish and shellfish, up from 55 percent in 1986. The PRC is probably the only country in the world where culture production exceeds capture production.

Aquaculture has also spread rapidly in other Asian economies, notably India, Indonesia, Japan, Republic of Korea, Philippines, and Thailand. However, in many of these countries, the commercialization and consequent intensification of aquaculture, the use of carnivorous species that depend on fishmeal extracted from capture fisheries, and the negative environmental and socioeconomic impact, have raised many doubts about its overall benefits and sustainability.

Asia has a long history of capture and culture fishery production, evident in the wide variety of traditional gear and culturing techniques that have evolved over time to exploit the diversity of resources. Fish form an important part of the diet of many Asians, although per capita consumption varies widely from country to country. It is generally low in land-locked countries, such as Bhutan, Kyrgyz Republic, Mongolia, Nepal,and Tajikistan, and in South Asia, but relatively high in Southeast and East Asia, particularly among wealthier economies.

The contribution of this sector to food security is highlighted by the fact that increased supplies of fishery products in the PRC raised the annual consumption of aquatic products there from 2.67 kg to 7.29 kg per capita during 1952 to 1992. This is especially significant given that the PRC population more than doubled in that period from 575 million to 1,172 million. The fishery sector in Asia employs a large workforce. Even though the proportion of people dependent on fisheries in Asia might appear small against the region's vast population, their number is considerable in absolute terms.

According to FAO some 25 million fishers and fish farmers-four fifths of the world total-are employed in Asia. In South and Southeast Asia, fisheries employ 10.36 million people as full- or part-time fishers, with 8.64 million employed in marine fisheries and the remaining 1.72 million in inland fisheries. Moreover, there may be a large number of occasional fishers, particularly in the inland fisheries. In addition to the direct employment provided by fisheries, considerable job opportunities exist in the related service and transport industries. Opportunities for women exist especially in aquaculture, fish retailing, and processing. In the PRC alone, the population engaged in these fishery-related activities (not including capture fisheries and aquaculture directly) numbers over 11 million.

Marine capture fisheries, which contribute the largest, albeit declining, share of Asia's total fisheries production, are characterized by the presence of a large number of small- and medium-scale fishers. They operate in shallow inshore waters of up to 50 m in depth, using traditional but increasingly modernized craft and gear as well as nontraditional craft and gear. Fishing pressure is intense along the continental shelves of the western coast of India, the Bay of Bengal, the Gulf of Thailand, the South and East China seas,the Bohai and Yellow seas, and the parts of the Sea of Japan bordering Japan.

The gross statistics of Asian fisheries might appear to contradict the general picture of ill health in most major fisheries of the world. Many of the world's major fisheries are facing serious falls in production; stocks of a number of commercially important fish species have been fully or overexploited or are rapidly dwindling, and many commercial species are endangered. The trend to fish "down the food chain" may have long-term and perhaps irreversible impact on the ecological balance in marine ecosystems.

The openaccess nature of fisheries and "subsidy-driven overcapitalization" have been largely responsible for the trends of overfishing and excess capacity that have led to a global crisis in fisheries. These trends already exist in many Asian waters, casting shadows on the sustainability of fisheries growth in Asia. The predominance of a single country such as the PRC makes sustainability of Asian fisheries growth even more vulnerable because it hinges largely on the ability of the PRC to sustain its high production rates, particularly in aquaculture.

However, even though further potential for aquaculture expansion in the PRC has been identified, the two major constraints for aquaculture generally, viz. environmental degradation (from aquaculture itself as well as from external sources), and competition for land and water resources from other economic activities, are also likely to threaten the goal of realizing that potential. Intensification of aquaculture has been contributing to further environmental degradation, which is already apparent in the PRC and elsewhere in Asia.

There are several indicators that suggest that overfishing in many parts of Asia may have worsened during the last two decades. Demersal stocks have been heavily fished in much of the Gulf of Thailand since the 1970s. In Thailand, catch rates in terms of catch per unit effort are currently only 6 to 10 percent of their peak levels, which were reached soon after the introduction of otterboard trawling in the early 1960s. Catches of a number of large and small pelagic stocks also appear to have declined, although the recent decline in landings of small pelagic fish is largely attributed to the environment-linked fluctuations in catches of the Japanese pilchard in the northwestern Pacific. More importantly, however, catches of miscellaneous species, which traditionally form a large part (nearly a quarter) of Asia's marine capture fisheries, have been on the rise.

Thus, 60 percent of the world's fisheries (including some in Asia) are in urgent need of management to control and reduce fishing capacity and effort. For instance, large increases in the capture of marine cephalopods and other molluscs are attributed to a decline in demersal fish, which are their predators. Catches of these molluscs may increase in the short term, since the prey population is much larger than that of the predators, but the value per unit catch decreases and the increasing catches give "a misleading vision of the state of world fishery resources and a false sense of security".

The declining share of marine capture fisheries to total fishery production in Asia is mostly attributable to slower growth in the northwestern Pacific, traditionally dominated by the commercial fisheries of PRC, Japan, Republic of Korea, Democratic People's Republic of Korea, and nonAsian countries such as Russia, the USA, and several European nations. Yet, the average production growth rate of 1.9 percent per year in this fishing area during 1987 to 1996 was still higher than the world average of 1 percent. The drastic decline in recent years in the catches of the two dominant species, viz., the Japanese pilchard (sardine) and Alaskan pollock, has been the main reason for the slower growth of the northwestern Pacific capture fisheries.

Marine capture fisheries still account for the largest share of Asian as well as global fish production. The share of inland capture fisheries has been marginal. These fisheries have grown at twice the rate of marine capture production during the past decade, in part due to the large number of reservoirs constructed and increasing efforts to seed inland waterbodies. However, further expansion may be constrained by environmental degradation and the fact that most inland waterbodies have been already exploited.

The fishery industry, dominated by capture production, is among the few remaining frontiers of hunting and gathering in human society. The pressure exerted on the world's fishery resources, owing to modern fishing technology and lax resource management, has been so intense that in most cases making fisheries sustainable would require control of access and reduction of fishing effort on overexploited resources, including demersals and straddling stocks (i.e. of species whose range includes the waters of two or more countries). These measures would have to continue until some recovery, indicated by increased catch rates, became evident. Management efforts should therefore focus on sustaining rather than increasing catches from marine fisheries. Largely due to population pressure on inland water resources, inland capture fisheries in Asia have not developed on a large scale, contributing only about 7 percent of total fish and shellfish production. Although inland capture production in Asia grew at the moderate rate of 4 percent during the last decade, its share in total production declined slightly.

Worldwide, inland capture production has stabilized at around 6.5 million t after peaking in 1990, and this level is expected to be maintained until at least 2010, although wide local variations may occur. The major constraints to inland capture fisheries are the growing threat from pollution due to increasing urbanization and industrialization, and inadequate access and user rights. It should be noted, however, that a significant part of inland capture landings does not enter into official statistics, because some of the harvest is consumed directly by fishers and their families. In Thailand, for example, such direct consumption is estimated at 25 percent of the reported catch.

Realizing the importance of inland fisheries (both capture and culture) to rural food security, intensive aquaculture and culture-enhancement techniques are being used in many Asian countries to increase productivity through seeding or stocking of waterbodies. With the further development of hatchery technology, these practices should offer greater potential for realizing higher yields. Overall, culture-enhanced capture fisheries seem to offer better potential for low-income, resourcepoor communities, because they use existing water resources and low resource-input systems, and create little, if any, pollution. They may thus be better suited than intensive aquaculture for rural communities.

Aquaculture

Long histories of aquaculture practices are also evident in a number of other Asian countries, such as polyculture of freshwater fishes in natural and human-made ponds and tanks in India and the tambak culture system in Indonesia. These traditional systems are mainly extensive forms of culture with little, if any, external inputs such as feed. The rapid spread of aquaculture and its diversification in many Asian countries is due largely to successes in hatchery technology and balanced-diet feed manufacturing. These and various other technologies have allowed intensification of aquaculture, increasing productivity.

Today, Asia is the largest producer of cultured fish, contributing almost 90 percent of the world culture production of 25 million t in 1994. Particularly high growth rates have been attained for freshwater finfish culture (carps, tilapia, and a number of other species) and the culture in marine waters of cephalopods (squids, octopuses, cuttlefish) and oysters, for which technological breakthroughs have been achieved and adopted all over Asia. Marine aquaculture has also grown at a fast pace. Its share of aquaculture rose from 33 to 37.5 percent during 1987-1996, while that of freshwater culture fell from 62 to 60 percent. In terms of value, however, brackishwater culture had a larger share (21 percent on average during 1987-1996), than its small volume (7 percent) would otherwise indicate. Although brackishwater culture overall grew relatively slowly, coastal aquaculture, particularly of penaeid shrimp, but also of cephalopods, grew at a high rate, especially for several major exporters such as Bangladesh, PRC, India, Indonesia, Thailand, and Viet Nam. In contrast, inland brackishwater culture declined in some countries, such as the Philippines and Thailand, and grew rapidly in others, particularly Bangladesh and Taipei,China.

The fast growth of aquaculture is a response to both highand low-end markets. Coastal shrimp culture has developed in response to increasing demand from international highincome markets and the resultant price increases, whereasfreshwater finfish aquaculture is usually focused on low-value food fish. For the former, the income earned has been illusory and deceptive because neither the farmers nor the governments have fully considered the overall cost of shrimp farming. The boom in freshwater low-value food-fish aquaculture, by contrast, has benefited the poor, because fish is a major source of protein for a large number of Asian countries, including Bangladesh, PRC, Indonesia, Myanmar, and Thailand.The growth of freshwater aquaculture in these countries has substantially raised the nutritional level of the poor.

After three decades of annual growth in fertilizer use of around 10 percent, Asia's croplands in 1996 received on average around 135 kg/ha of fertilizer, from 38 kg/ha in 1965. In irrigated areas, fertilizer consumption per hectare is much higher than the national average. Malaysia, PRC, Republic of Korea, and Viet Nam use over 700 kg/ha of fertilizer on irrigated land (Table 4). In Indonesia, the rate is over 500 kg/ha. Countries with low consumption levels, such as Cambodia and Myanmar, tend to be those with foreign exchange problems, lack of proper distribution and credit systems, and lack of incentives resulting from price ratios of grains to fertilizers. Much higher rates are applied in intensive cropping systems.

Table 4: Total Fertilizer Consumption in Irrigated Areas in Asia

	Total Fertilizer Consumption (kg/ha)			Annual Growth Rate (%)	
	1975	1985	1995	1975–1985	1985–1995
East Asia					
China	160.17	378.00	713.67	9	6
Japan	568.05	689.02	609.26	2	-1
Korea	677.76	609.06	714.73	-1	2
Mongolia	152.17	310.00	31.25	7	-23
Southeast Asia					

Cambodia	1.12	0.00	58.38	-1	41
Indonesia	125.41	458.56	558.17	13	2
Lao PDR	2.50	16.81	34.97	19	7
Malaysia	805.26	1,830.54	3,323.53	8	6
Myanmar	56.07	178.88	109.59	12	-5
Philippines	218.17	196.65	381.72	-1	7
Thailand	74.47	113.44	311.64	4	10
Viet Nam	330.00	217.85	724.00	-4	12
South Asia					
Afghanistan	14.97	28.21	17.86	6	-5
Bangladesh	149.49	260.82	372.59	6	4
Bhutan	4.55	3.33	2.56	-3	-3
India	103.58	203.55	276.97	7	3
Nepal	53.30	57.12	105.90	1	6
Pakistan	40.63	95.88	145.80	9	4
Sri Lanka	150.83	335.31	363.38	8	1

0 = zero or less than half of the unit measured.

Note: Annual Growth Rate = ((Ln(value year 1) - Ln(value year 2)) / number of years) x 100.

Indications of inefficiencies in fertilizer use are much less obvious than in insecticide use. No major ecological or economicdisasters on the scale of the brown planthopper epidemics have yet been reported. Many authors have, nevertheless, pointed to two possible types of impact of fertilizer use in terms of (a) fertilizer-use efficiency and (b) nutrient imbalances. Declines in the ratio of grain to fertilizer, e.g. in India from about 60:1 in 1966 to less than 10:1 in 1992 for rice and from 15:1 to 5:1 for wheat, have caused concern about a possible decline in the efficiency of fertilizer use. This has, however, been indicated as being somewhat misleading because the grain:fertilizer ratio is not a very accurate indicator of fertilizeruse efficiency; it erroneously assumes a zero yield in the absence of fertilizer use.

In Karnal (in Haryana, the heart of India's green-revolution territory), the marginal response to fertilizer (the ratio of increase in yield to increase in fertilizer) for rice has declined somewhat. This can probably be explained by the diminishing returns from the very high rates of fertilizer now being applied. For wheat in India, the marginal response was still increasing slightly in the early 1990s. It is now technically quite simple to achieve the twin goals of improving fertilizer-use efficiency and soil nutrient levels. Losses from nitrogen fertilizers can be effectively minimized through the use of such innovations as urea supergranules or deep placement of urea, and urease inhibitors. Fertilizers that are well balanced in relation to crop requirements and the soil's own nutrient capacity can be easily formulated with help from soil and plant analysis and fertilizer trials. Most of these, however, are still not yet applicable for use on the average Asian farm.

Most farmers consider the deep placement of urea as too labor intensive. Supergranules are very costly. The urease inhibitors, which were only in the experimental stage in the early 1990s, have not yet been incorporated into the fertilizer manufacturing process. Access to services that would help to improve the match between the nutritional content of fertilizer, the capacity of the soil to supply nutrients, and the needs of crops is still unavailable to most Asian farmers. In developed countries, individual farmers sometimes use tools such as plant and soil analysis and fertilizer trials for fertilizer management. More often, however, these facilities are provided as part of the service rendered by fertilizer companies and farm consultants. Such services are rare in Asia. Among the exceptions are the consultancy services that provide advice to the larger oil palm and rubber plantations in Malaysia and Indonesia. Their fertilizer recommendations are generally based on tissue analysis.

Some fertilizer companies that provide soil analysis services as part of their marketing operations are now found in the region. Lack of analytical facilities is not the main reason for the lack cf services to support improved fertilizer management in Asia. Most analytical laboratories, many exceptionally well equipped by development assistance programs, are actually greatly underutilized. Logistical arrangements are lacking on how to take samples, determining where to send them, making sure of the timely return of results, and interpreting the results. The services provided by the analytical laboratories are, therefore, of little use to district farm advisors, farmers, or fertilizer marketing personnel. Most analytical chemists would also point out that the results from plant and soil analysis are useless without stringent quality control on laboratory procedures.

Most farmers in Asia have few resources for fertilizer management other than the "official recommendations". These recommendations are not generally very responsive to local variations or the impact of cropping intensification that has taken place in the last 20 to 30 years. For example, despite the thousands of fertilizer experiments that have been conducted in the past two decades throughout India and Pakistan, practically the same fertilizer recommendation is given in all irrigated areas. Some experts are now advocating alternative agriculture, defined as alternatives to high-input technology such as that of the green revolution.

Irrigation

Irrigation was key to the success of the green revolution. Irrigation not only augments the water supply but also improves and ensures the stability of water delivery, widens crop choices, and allows increased cropping intensity. Asia has 179 million ha or 69 percent (in 1995) of the world's irrigated areas. The PRC alone has 49 million ha under irrigation, India 50 million ha, and Pakistan 17.2 million ha. Over the past two decades, irrigated areas have increased in most Asian countries, especially Bangladesh, Bhutan, PRC, India, Pakistan, Thailand, and Viet Nam, but the rate of increase has slowed down in the last decade except in PRC, Myanmar, and Bangladesh. Since 1980, the irrigated area in Asia has expanded at the rate of about 2 percent per year; about 35 percent of the arable land in the region is now under irrigation. Future growth in irrigated area may come from India where it is planned to add 17.3 million ha of irrigated land by 2020 (Rosegrant and Ringler, 1998). The general reduction in growth of irrigation is a result of the decline in funding by major lending agencies as well as of the difficulties in finding projects with high returns. In many countries, such as PRC, Japan, Republic of Korea, and Sri Lanka, the supply of land that would yield a high return under irrigation has been mostly exhausted. In other countries, such as India and Thailand, where expansion is being planned, the marginal cost of irrigation is high when social and environmental costs are taken into account.

Since 1980, the efficiency of irrigation systems has become an issue of increasing concern. Poor maintenance and rapid deterioration are common features of irrigation systems in many Asian countries. Irrigation agencies are interested in increasing physical capacity without commensurate increases in management capacity. Planned capacities have fallen short ofactual needs, and some systems are unused owing to lack of water, inappropriate design, or poor maintenance. The overall system efficiency is low, for example 30 and 38 percent in northern India and Karnataka, respectively.

In Central Asia, the breakdown of the drainage system in salt-affected irrigated areas has led to further elevation of the groundwater table, thereby increasing salinity, which has led to large yield losses and finally to the total loss of cropland in some areas. Communal irrigation systems that have been taken over by centralized irrigation agencies often become inefficient because of a lack of appreciation of onfarm water needs. A study of 15 irrigation systems in South and Southeast Asia indicated the lack of two-way communication between irrigation agencies and water users.

At the farm level, farmers tend to use more water than is needed. For rice, the amount of water used may be 6 to 10 times more than is necessary. Water pricing has been suggested as a means of overcoming water waste, but farmers will then have to weigh the cost of water with the cost of weed control. Another method for saving water is the conversion from transplanting to direct seeding which reduces water use by half, although the yield may be lower even with good weed control. Other irrigation techniques, e.g. drip irrigation and methods that are site-specific water applications, are emerging where water is scarce and the crops are of high value.

In the past, the construction of multipurpose or agricultural large-scale dams was often planned top-down, with insufficient consideration given to the people who would be affected by the project. The impact on forests resources and biodiversity was usually not taken into account. Recently, some well-organized NGO networks have effectively publicized the plight of dam refugees and the ecological costs of large-scale infrastructure projects. This has rendered projects in the region more transparent and has enhanced the accountability and worthiness of the projects.

The process for determining economic feasibility and environmental impact needs to be strengthened to improve irrigation efficiency and to avoid having dams that have insufficient inflows or that lead to an increase in soil salinity. To date, the feasibility study and the environmental impact assessment have been important only in terms of the loan application process. Irrigation projects are often not transparent or accountable to the public. Cost-benefit analyses have not been rigorously conducted. It is important that in the future, projects should involve the participation of a wider spectrum of stakeholders.

Recently, there has been global recognition of the value of consulting and involving water users in water management plans and activities related to irrigation systems. For the past two decades, more and more countries around the world have been turning over management authority for irrigation systems to farmers' groups or local entities, in a process commonly referred to as irrigation management transfer (IMT). There have been several studies on this process and the literature shows a mixture of positive and negative results. Although most of the studies are deficient in assessing the real cost of farmers' participation, government expenditures for irrigation tend to decline and costs to farmers often rise. There is little evidence to suggest that yield, water productivity, or farm income has increased.

Poor operation and management have a negligible impact on the irrigated crop. Studies that would make it possible to separate the impact of IMT from other factors such as weather are lacking. In many instances, the responsibility for rehabilitation is not clearly spelled out in the IMT agreement between the government and local entities. The key to sustained success of farmers' participation is the incentive structure and quality of leadership, which can vary widely from place to place and from time to time. There is no available model to follow for molding the farmer-agency relationship that will work in all societies in all situations. Many innovations may be needed to develop the right relationship for a given set of conditions. It is hoped that as the real value of water becomes better understood by all users and as more realistic water pricing becomes feasible, workable models will emerge for sharing responsibility between agencies and users in managing irrigation water.

CHAPTER 6

TECHNOLOGY IN CROP IMPROVEMENT

IN VITRO CULTURE FOR MICROPROPAGATION

Vitroplants of potato, sweet potato, Allium species, taro, strawberry, sugar cane, several ornamentals such as chrysanthemum, carnation, Lilium species and gerbera, and fruit plants such as Citrus species, papaya, apple, sweet cherry, pear and grapes freed of or free from viruses are being commercially exploited in several countries in the region, particularly in China, India, Indonesia, Japan, Malaysia, the Philippines, the Republic of Korea and Thailand. Hundreds of thousands of hectares are planted to virus-free vitroplants of potato in the Region with more than doubled productivity, especially in China and the Republic of Korea. For instance, in the Republic of Korea, the widespread use of virus-free planting material produced through in vitro culture techniques increased the national potato yield from 11.9 tonnes/ha in 1980 to 20.3 tonnes/ha in 1986.

Subsequently, an in vitro tuberization system was established and became an integral component of the potato seed industry in the country. Production efficiency of the technique is exceptionally high. One of the main problems encountered in this approach is the chance of reinfection during the propagation of virus-free plants. To overcome and manage this problem, simplified diagnostic methods have been developed and are being used to detect viruses and viroids. The vitroplants are being routinely mass micropropagated for commercial purposes in the case of orchids, other flowers and ornamental plants throughout the region, especially in Southeast Asian countries, oil palm (Malaysia, India, Indonesia, Papua New Guinea), date palm (Iran and likely to be taken up in India and Pakistan), embryo culture of soft coconuts (Macapuno) (the Philippines and Thailand), cardamom (India; Kumar, 1990), eucalyptus (China and India), Chinese fir (China), rattan (Indonesia, Malaysia, the Philippines), and seed sterile triploid water melons (Japan).

Somatic embryogenesis is also being commercially applied. Synthetic seeds of rice and vegetables have been developed by the private sector in Japan. Triploid rubber clones of rubber produced through somatic embryogenesis have outyielded diploid standard clones by about 20 percent in China. The country has also standardized techniques for mass propagation of sugar cane seedlings using embryogenic cell lines and multishoot propagation systems. To facilitate reforestation and promote social forestry and agroforestry, India has established two tissue culture pilot plants, one at the National Chemical Laboratory, Pune, and the other at the Tata Energy Research Institute, New Delhi (a private organization) with production capacities of five to ten million propagules/seedlings of elite/plus trees of several important species, such as Eucalyptus tereticornis, E. camadulensis, Tectona grandis, Dendrocolamus strictus, Populus deltoides, Bambusa vulgaris and B. tulda. Micropropagation of teak, rattan, eucalyptus, bamboo and other tree species has been adopted in several other countries of the region.

Haploid Breeding Methods

The region has played a pioneering role in this field. Indian scientists were first to produce haploids from another culture in the mid-1960s. Haploid induction through anther culture has been used most extensively by Chinese scientists for breeding of not only rice but also of several other crops. The technique was used for: (i) production of new varieties; (ii) production of inbreds for heterosis breeding; (iii) selection of stable resistant lines against biotic and abiotic stresses; and (iv) purification of male sterile lines. One of the most important advantages of the use of haploids is to reduce the time required to develop and release new varieties. For instance, while the conventional pedigree method of breeding and releasing a pure line rice variety requires about 12 years, the haploid method requires only about five years.

Chinese scientists have succeeded in inducing haploids in about 40 different plant species. They were the first to produce haploids through in vitro culture in wheat, maize, sugar cane, soybean, rubber, grapes and apple. In China, several new varieties of rice, wheat, tobacco, and hot and sweet pepper possessing high yield, superior quality, sometimes tolerance to abiotic stresses such as cold, sometimes early maturity, and resistance to diseases have been released through the use of haploids. The most popular of these varieties are Xin-Xin, Hua-Hau-Zao, Zhong-Hua 8, 9, 10 and 11 of rice; Jing-Hua 1, 3 and 5 of wheat; Haihua 1, 3, 19 and 29 of hot and sweet pepper; and Dan-Yu 1, 2 and 3 of tobacco. These varieties occupied about 1.5 million hectares in 1990, of which the rice and wheat varieties accounted for 800 000 and 650 000 ha, respectively.

Other countries have also used doubled haploid breeding methods for crop improvement. In India, rice lines R4 and R10 derived through this method outyielded the check by 15 to 49 percent in trials conducted during 1984-87. The Republic of Korea released two rice varieties derived from anther culture haploid technique and generally screens annually about 6 000 anther-derived rice lines. The technique has also been effective in heterosis breeding of Chinese cabbage. In Japan, several successful varieties have been developed through this technique. Among these are rice varieties Joiku N. 394, Hirohikari, Hirohonami, AC No. 1 and Kibinohana, which are tolerant to cold temperature and are good in taste. A broccoli variety, "Three Main", possessing cold resistance and uniform shape, and a cabbage variety, "Orange Queen", with an orange colour, which are quite popular, were developed by the anther culture method.

Somaclonal Variation

Australian scientists were the leaders in the field of somaclonal variation, demonstrating the efficacy of the approach in improvement of wheat, sugar cane and other crops. Somaclonal variants and lines selected through selection pressure exerted during the culture stage have been released as commercial varieties in several countries in the region. For instance, in Japan, the 1991 list of registered crop varieties included two varieties of Lilium and one each of melon, Gerbera, Cymbidium, statice and strawberry derived through somaclonal variation. In China, somaclonal mutants of rice possessing resistance to bacterial blight and resistance to AEC coupled with higher contents of Iysine, methionine, isoleucine, serine and tyrosine than the parental lines have been produced.

Salt tolerant somaclones of rice, wheat and tobacco were also isolated. Short stature and early maturing somaclonal variants of rice have been isolated in several laboratories in China. India and other countries have also produced somaclonal variants and are using them directly as new lines or in breeding programmes. In Fiji, tissue culture techniques have been used extensively for isolating sugar cane clones resistant to Fiji disease of sugar cane. Thailand isolated improved varieties of banana and chrysanthemum using this technique.

Somaclonal variation was exploited in a novel way under a joint programme of China and Australia to transfer from Thinopyrum intermedium resistance to Barley Yellow Dwarf Virus (BYDV) in wheat (Triticum aestivum). Single cell callus cultures from F1 embryos rescued were initiated and induced to form plants showing somaclonal variation which were then selected for BYDV resistance. Cytological analysis of the genotypes showing stable resistance revealed that chromosomal rearrangement of the chromosome carrying Thinopyrum introgressed segment had occurred during the tissue culture phase to confer the stability.

The Biotechnology Centre at the Indian Agricultural Research Institute (IARI) has standardized the protocols of plant regeneration of Brassica carinata and is isolating somaclonal variants suitable for Indian conditions. Useful somaclonal variants for earliness, plant height, maturity, etc. have been induced in B. juncea and B. napus. In vitro techniques for plants provide systems analogous to the prokaryotic systems where variants can be selected and mutations can efficiently be induced and isolated at cellular level.

By applying effective selection pressure on naturally variant or mutagenically treated cell cultures viz. use of saline media for screening salinity resistant/tolerant cell lines or use of pathogen toxins for isolating disease resistant cell lines, the efficacy and efficiency of both induction and identification of useful mutants are increased considerably. Using this technique, Chinese, Indian and Filipino scientists have isolated rice mutants tolerant to higher concentrations of salt. Mutants having higher protein and Iysine content in their seeds were also isolated. Using this approach, pathogen-resistant mutants of tobacco, rice, wheat, barley, sugar cane and maize have been isolated and used in breeding programmes in several countries. In Japan, disease-resistant lines of rice, tomato and tobacco were isolated using this approach (Table 1).

Table 1: Examples of disease-resistant plants successfully selected using tissue culture, Japan

Crop	Disease	Selection stress	Selection stage
Oryza sativa	Rice blast	Tenuazoic acid	Protoplast-derived callus
Lycopersicon esculentum	Tomato fusarium wilt	Fusaric acid	Leaf-derived callus

Avena sativa	Oat leaf stripe	Victorin	Regenerants from callus and their progeny
Nicotiana tabacum	Tobacco mosaic virus	No stress	Shoots regenerated from TMV-infected callus
Lycopersicon esculentum	Tomato bacterial wilt	NO stress	Regenerants from callus

Embryo Culture

The technique is being used in several Asian countries, particularly in the case of orchids, peanuts, cotton, cabbages, citrus and peaches. In India, promising recombinants have been obtained using embryo rescue techniques in distant crosses of cultivated Phaseolus, jute and peanut. In China, using embryo rescue and in vitro culture of a distant hybrid Triticum aestivum x Agropyron elongatum, a new wheat variety, Xiaoyan No. 6, was produced and has been grown widely giving additional wheat production. In Japan, three cultivars each of Citrus, Prunus, and Brassica species, and five cultivars of Lilium species were developed in recent years using the embryo rescue technique.

Somatic Hybridization

Japanese scientists have reported successful protoplast culture in more than 70 plant species. In China, plants regenerated from protoplast have been obtained in about 30 species, including vegetables, medicinal plants, legumes, and other economic crops as well as woody species such as poplar, elm and rubber trees. For the first time, Chinese scientists regenerated plants from monocot Polypogen fugax. With recent reports of success on protoplast culture and regeneration of wheat, it is now possible to have protoclones of most of the major food crops. However, in several cases, the regeneration frequency is low and should be improved. The Republic of Korea had produced cybrid lines of mushrooms which outyielded the best parents and checks by about 100 percent.

Successful regeneration of plants from cell fusion are reported from about 50 interspecific and intergeneric protoplast combinations in the region. Half of these are known from Japan. Using cell fusion techniques, scientists have developed novel citrus hybrid varieties, such as "Oretachi" (orange plus trifoliate orange), "Shuvel" (Satsuma mandarin plus navel orange), "Gravel" (grapefruit plus navel orange), "Murrel" (murcott plus navel orange) and "Yuvel" (Yuzu plus navel orange). Using asymmetric cell fusion techniques, Japanese scientists developed Ms-F 224, the first commercial tobacco male sterile line developed in the world bred by cell fusion. Such asymmetric male sterile lines for commercial exploitation of F_1 hybrids have been also bred in carrots, cabbages and eggplants.

In China, somatic hybrids between Nicotiana tabacum and N. rustica, N. tabacum and N. glauca were used for developing new commercial varieties of tobacco. Chinese scientists have also pioneered the pollen tube gene introduction technique for cotton and rice breeding. In this technique, after self-pollination, the desired exogenous DNA is introduced to the embryonic cells. Seeds which develop from such transformed embryos result into transformed plants, thus avoiding the need for protoplast fusion technique. Genes responsible for disease resistance and other traits have been transformed successfully in rice and cotton.

But this technique has low repeatability. In fact, the entire cell fusion technique has not been as successful in producing somatic hybrids as initially expected. With the availability of Agrobacterium-mediated and biolistics-based methods of gene transfer, the thrust on the protoplast fusion approach has somewhat slackened, although the asymmetric fusion method has special appeal for production and diversification of cytoplasmic male sterility and manipulation of other cytoplasmically controlled systems.

Secondary Metabolites

Cultured plant cells retain their metabolic potential and can be used efficiently for producing useful secondary metabolites of commerce such as pharmaceuticals, dyes, food additives, sweeteners, flavours and taste enhancers, essential oils and aromatic products. Herbal medicines are popular in China, India and other Asian countries and are gaining importance in the West. Chinese scientists have been using tissue culture since the late 1970s for the production of ginseng sapanins from Panax ginseng. Diasgenin, a female contraceptive produced from in vitro cultured Dioscorea spp and a male contraceptive based on gossypol from tissue-cultured Gossypium spp are under extensive trials in China. In addition, tissue culture technique has been successfully used in the cultivation of Scopolia acutangula, Artemisia annua, and Rauwolfia yunnanesis. India is also using this technique for production of diasgenin and other medicinal and aromatic products.

Cell cultures can also be used as factories for bioconversion of intermediate compounds into more valuable products. Shikonin, an expensive compound obtained from the roots of Lithospermum erythrorhizon, has been used by the Japanese traditionally as a vegetable dye and in cosmetics and toiletries. However, due to overexploitation and other human activities, the plant has become almost extinct in Japan. To reduce their dependence on the import of this plant material, Japanese scientists have developed a tissue culture method for the commercial production of shikonin. Another example where tissue culture production of an industrial compound has reached commercial level is berberine from Coptis juponica.

In tissue cultures the yield of high value compounds can be enhanced by feeding the cells with the intermediate compounds of their biosynthetic pathways (biotransformation), manipulation of culture conditions and selecting high yielding cell lines. With further refinements of techniques the bioreactor and fermenter based production of secondary metabolites could be rendered highly cost-effective and time-saving.

IN VITRO CONSERVATION OF GERMPLASM

Several plant species in Asia, including a few commercial species, produce recalcitrant seeds and it is thus difficult to conserve them through seeds. Furthermore, some species are shy seed bearers and even fail to produce seeds. In addition, living collections of clonally propagated perennial crops face the problems of maintenance of heterozygous and heterogeneous populations, the long life cycle and large space required, and a high possibility of exposure to the threats of pests and diseases and other biotic and abiotic stresses. To circumvent these difficulties, in vitro conservation of vegetatively propagated and recalcitrant seed-producing species is being increasingly adopted as a complement to other methods of conservation. For short- to mediumterm storage (working and active collections), the slow growth method is used whereas for long-term storage (base collections) cryopreservation is the method adopted.

There is therefore a good case in the region for in vitro storage of several of the important plant species, and the countries involved are in fact building such facilities. Orchid germplasm is routinely maintained in vitro in India, the Philippines and Thailand. With assistance from the Department of Biotechnology, the National Bureau of Plant Genetic Resources (NBPGR), in India has established extensive in vitro storage facilities and is already storing germplasm collections of certain root and tuber crop species as well as a few fruit species. The bulk of the Indian potato collections at Central Potato Research Institute, Shimla, are conserved through tissue culture. China is also developing comprehensive facilities for in vitro conservation.

The Southeast Asian Banana Germplasm Bank, currently maintained as a living collection in Davao, the Philippines, has already been put under in vitro storage. The country is also maintaining some of its desired soft coconut germplasm through embryo culture. Malaysia is in the process of duplicating some of its ex situ living collections of oil palm and rubber as in vitro collections. Several of the Pacific Island countries are maintaining their taro collections in vitro because of the danger of the accessions being lost to viral and other diseases under field conditions. Thus, there is wide scope for using biotechnology for conservation and utilization of tropical plant germplasm.

Diagnostics

Besides the developed countries, several developing Asian countries are using monoclonal antibody techniques for identification of pathogens and for indexing materials. In China, more than 50 kinds of hybridoma strains have been constructed which secreted various kinds of monoclonal antibodies to viruses, bacteria, cancer, etc. India is applying monoclonal antibody techniques for the diagnosis and epidemiology of tungro virus of rice and other important viruses. The Central Potato Research Institute, Shimla, India, has been using monoclonal antibodies, cDNA and enzyme-linked immunosorbent assays (ELISA) for detecting and characterizing viruses of potatoes. Several other developing countries have also been using this technique coupled with ELISA tests to detect peanut viruses. Thailand has been using cDNA probes for detecting Tomato Yellow Leaf Curl Virus (TYLCV), Papaya Ringspot Virus (PRY) and mycoplasma causing sugar cane white leaf.

Recombinant DNA Technology

As regards the developed countries of the region, Australia and Japan have comprehensive programmes on recombinant DNA technology and production of transgenics for commercial exploitation. Australia was the first country in the world to have engineered and released a micro-organism for biological control of crown gall in plants. Australia's first field trial of transgenic plants took place in 1991, involving transgenic potato varieties resistant to leafroll virus, developed by the Commonwealth Scientific and Industrial Research Organization (CSIRO) Division of Plant Industry. During 1988-91, several laboratories in Japan, using electroporation or Ti or Ri, reported successful production of transgenics in Oryza saliva, Citrus sinensis, Cucumis melo, Lactuca saliva, Solanum tuberosum, Nicotiana tabacum, Morus alba, Actinidia chinensis, Atropa belledona. Brassica oleracea, Lycopersicon esculentum, Vigna angularis and Vitis vinifera.

For effective gene expression, promoter regions/genes have been identified for tissue-, age- and pathogen-specific expression of the genes under transfer. Stable and useful transgenics for protein quality and for viral resistance have been produced in rice, potato, tomato, melon and tobacco (Table 2).

Table 2: Transgenic plants of agricultural crops produced using useful genes, Japan

Crops	Genes transferred	Transformation system
Oryza sativa	Rice glutelin	E.P.
	RSV coat protein	E.P.
	Bar (bialaphos resistance)	E.P.
Solanum tuberosum	α-amylase	Ti
	Soybean glycinin	Ti
Lycopersicon esculentum	TMV coat protein	Ti
Cucumis melo	CMV coat protein	Ti
Nicotiana tabacum	cDNA of mild TMV strain	Ti
	ttr (wildfire disease resistance)	Ti

Molecular studies, gene tagging and mapping studies have been intensified in several laboratories in Japan. In 1991, a rice genome project, on the lines of the famous human genome project, was initiated. Restriction Fragment Length Polymorphs (RFLPs), YAC clones or cosomid library and Several Sequence-Tagged sites (STSs) approaches are being deployed for the purpose. Cloned chromosome segments including useful genes are being analysed and characterized at molecular level.

For identification and characterization of rice somatic chromosomes in in situ hybridization, imaging methods have been developed. Several countries in the region are using RFLP, RAPD and other techniques for gene tagging and creation of genome maps not only of major food crops but of topical industrial and fruit crops such as rubber, oil palm, durian and mangosteen. This approach is likely to increase the interaction between biotechnologists and plant breeders.

In China, protocols for production of transgenics for disease resistance in rice, tobacco, soybean and Brassica sp. have been standardized and transgenics are at various stages of development. Chinese scientists have attained considerable success in developing oilseed rape lines resistant to turnip mosaic virus through genetic engineering means. In India, transgenics are at various stages of production in cotton, rice, jute, Vigna aconitifolia and Brassica sp. In the Republic of Korea, transformants were produced in tobacco, rice and tomato and work has been intensified for identification and isolation of useful genes, especially for disease resistance. The country is developing improved vectors for increasing the efficiency of transformants production.

Malaysia is engaged in developing transgenics rubber and oilpalm, besides rice. In Pakistan, biotechnological manipulation of protein content and quality and resistance to Ascochyta blight in chickpea has been pursued in the last few years. Thailand, using the coat protein approach, has developed tomato and papaya lines resistant to TYLCV and PRV, respectively.

APPLICATION OF BIOTECHNOLOGY TECHNIQUES

The application of biotechnology techniques within the agriculture sector can potentially improve food security by raising crop tolerance to adverse weather and soil conditions, by enhancing adaptability of crops to different climates and by improving yields, pest resistance and nutrition, particularly of staple food crops. Genetic modification of crops is one such method allowing individual characters (gene, factor or trait) to be transferred into crop plants. With the advent of genetic modification through genetic engineering in early eighties, the natural barrier of only intra-specific exchange of characters was removed and scientists were able to identify and transfer specific genes associated with desirable traits from one organism to the organism of other species that otherwise cannot breed naturally.

With these techniques genes from varied class of organism like bacteria, viruses or even animal may be transferred into plants to develop genetically modified plants having exclusively changed characteristics controlled by the specific gene. This gives scientists/breeders a broader access to desirable traits from any living organism and its possibility of transferring it with much faster rate and greater precision. The Department has successfuliy built-up a strong base for taking advantage of this innovative science. Continuous efforts during 9th Plan have added strength to this mission. The Department has made concerted efforts for supporting various programmes in this area since 1990.

Presently, there are three types of ongoing research activities: R&D Projects on priority crops, Multi-institutional projects. The mandate of these programmes relate to major problems of priority crops, development of transgenics, quality improvement and basic research in the area of plant molecular biology. There are a number of multi-institutional projects e.g. in wheat programme was initiated for characterization of quality traits; in mustard- development of molecular methods for hybrid seed production; development of transgenic cotton, rice, mungbean and tomato for resistance to various biotic stresses; salinity and dehydration stress tolerance in rice; cloning of responsive genes and their promoters and development of wheat varieties with dial resistance to leaf and stripe rust using molecular marker technology.

The Department took the initiative to join the International Rice Genome Sequencing Programme (IRGSP) in June 2000 with a commitment to sequence 10 Mb of Chromosome 11 in five years. The programme is running ahead of time and India has fulfilled the international obligation in two and a half years time, till date a total sequence data of 15.80 Mb has been deposited.

Under the wheat network project different approaches were used to develop molecular markers for tagging genes with quality traits. However, PCR based approaches detected adequate polymorphism. STS and STMS markers linked to three traits (pre-harvest sprouting, protein content and grain size) were identified; also ISSR, RAPD and STMS markers linked to protein content and grain size were identified.

In another project on development of molecular methods using barnase and barstar genes for hybrid seed production, in mustard transformation of B. juncea cv Varuna with constructs has been initiated. A number of transgenics have been developed with different barstar constructs and plants were selected in vitro, on herbicide barstar. Constructs have been designed in which influence of 35 S promoter is limited and transgenics with barnase gene showing male sterile flowers have beenobtained.

Insect resistant transgenic rice developed through introduction of a modified endotoxin gene of Bt. Further bioassay of putative transgenics and their progeny for expression of Bt gene is being carried out. Molecular constructs have been made both for insect and viral resistance. Transgenic technology has been standardized and found to be most amenable to cotton and rice. Transgenic system of indica rice and wheat has been developed. Rice has been transformed with codA/COR47,

AtHSP100 and PDC gene, while wheat was transformed with HVA1 to confer stress resistance. All the transgenics have been characterized at the molecular level for gene integration.

For developing Vigna mungo plants with yellow mosaic virus resistance, coat protein gene in sense orientation and replicas gene in both sense and antisense orientation have been cloned. The binary vector has been mobilized into A. tumifaciens. Transformation of V. mungo and selection of transformed plants is in progress. Studies have been conducted to transform IR-50 rice plants with P5CS gene to make it more tolerant to salinity. The assessment of transformed plants harboring this gene in single copy and double copies showed promising results.

For incorporation of synthetic Bt Cry 1A(b) gene into vegetable Brassicas, the presence of transgene was checked via PCR. Southern hybridization was also done. Individual transgenics were found to carry 1-4 copies of Cry 1 A(b) gene. Northern and Western analysis confirmed the expression of transgenes.

For tagging genes for LF resistance in rice, two RAPD markers linked to a major gene of leaf folder resistance have been identified. Based on these markers SCAR primers have been developed.

For tagging genes for brown plant hopper resistance in rice, several RAPD and SSR markers for BPH resistance have been identified. Also conversion of a dominant RAPD marker AJ9 to a Co-dominant STS marker done and this marker is used in the selection of genotypes for BPH resistance.

Two salt inducible genes, PC kinase and proline serine rich protein encoding genes (PCSDR) were isolated from cDNA library of Porteresia coarctata. Both the genes have been characterised at nucleotide level.

Interesting leads obtained was on salt tolerant inositol synthase gene from halophytic rice, Porteresia coarctata.

The reproducible protocol for somatic embryogenesis and direct shoot regeneration has been established for three popular sugarcane varieties viz.

After standardization of transgenic technology, now emphasis is on development of large number of transgenics in cotton, rice, mungbean and tomato followed by diversification, molecular characterization and field evaluation. The international obligation of Rice genome sequencing data was fulfilled by the end of December 2002. Under the collaborative programme, transgenic Brassica is being tested for the male sterility, herbicide resistance and Alternaria blight in the field conditions.

Transgenic system in indica rice and wheat have been developed. Rice has been transformed with codA/COR47, AtHSP100 and PDC gene while wheat transformed with HVA1 to confer stress resistance. All the transgenics have been characterized at the molecular level for gene integration. Studies have been conducted to transform indica rice plants with P5CS gene to make it more tolerant to salinity. The assessment of transformed plants harboring this gene as a single copy showed promising results and further efforts are on to pyramid one more gene for higher tolerance to salinity.

The application of biotechnology techniques in the agriculture sector can potentially improve food and nutritional security by raising crops tolerant to adverse weather and soil conditions, by enhancing adaptability of crops to different climates and by improving yields, pest resistance and nutrition, particularly of staple food crops. Over the past decade, the application of biotechnology to the problems in world agriculture has yielded significant productivity gains to the producers. Various programmes of the Department have been directed in this direction. With advancements in GM technologies, these benefits are expected to increase in Indian scenario.

10th Plan Initiatives

— The rice genome-sequencing programme with present

commitment of gap filling and annotation shall be given high priority.

— A collaborative network programme on functional genomics of rice would also be considered. It has been decided to support one collaborative multi-institutional project each in the area of 'strategic research' and another on 'impact research and delivery'. The following two projects are under consideration (i) gene expression profiling during flower and seed development and functional validation of identified genes and (ii) identification and functional validation of genes related to yield and biotic stresses.

— It is intended to start a Wheat Biotechnology network with an aim to sequence wheat genome and mapping of important traits.

— A collaborative programme on the use of biotechnology for improvement of semi- arid crops like Sorghum and Chickpea in association with ICRISAT and National Research Laboratories would be initiated.

— A network project on the use of molecular marker technology and genetic engineering approach in wheat quality breeding shall be launched on the basis of leads obtained from previous project.

— As a follow-up action of Indo-US workshop on agricultural biotechnology, it is proposed to initiate a major multi-institutional programme on abiotic stress tolerance in rice, chickpea and wheat.

— The application of molecular biology tools in crop breeding with special reference to improvement of quality, protection from biotic/abiotic stresses will receive high priority.

— Exploitation of hybrid vigour by developing and using genetically engineered male sterility in selected crops.

— A programme on engineering the biosynthetic pathway of starch in rice and wheat would be considered.

— Development of marker assisted selection techniques for use in breeding for tolerance to drought, submergence and salinity in selected crops.

— The ongoing efforts using molecular biology tools for protection against biotic stresses will receive high priority. The list of target crops shall be increased with the perfection of technology.

— For combating malnutrition, efforts would be made to transform high yielding varieties of staple food crops with enhanced protein contents, development of rice varieties with p- carotene and transformation of rice and wheat for high iron content.

— Engineered male sterility for extended hybrid technology to more crops shall be exploited.

— Major genes representing Quantitative Trait Loci for tolerance to abiotic stresses (drought, moisture, salinity, heat, heavy metals and toxic substances), resistance to biotic stresses (viruses, bacteria, fungi, insects, nematodes), plant architecture (height, leaf area and thickness, root profile), flowering time, yield and mineral nutrition would be the main targets of study and developing transgenic crops..

— Fine mapping of genome regions harboring useful genes would be done by taking advantage of available sequence data of genomes to tag genes for marker-assisted precision breeding.

Crop Transformation

In the tropics, average crop production is reduced by 11-25% as a result of damage by nematodes. Crop resistance is a low-cost option for nematode control in subsistence agriculture. It does not impose unwanted changes on traditional agronomic practices. The range of single traits available for breeding programmes currently limits the number of nematode-resistant crops available. In addition, pathotypes, or species that overcome resistance, often challenge the utility of resistance.

The use of transgenes broadens crop resistance by protecting plants from nematodes for which natural resistance genes are unavailable. Broadly based nematode resistance also avoids the need for farmers to recognise and distinguish nematode species. Genetically engineered resistance to rice yellow mottle virus. Rice yellow mottle disease, caused by rice yellow mottle virus (RYMV), is a major limiting factor in African rice production. Rice yellow mottle disease is increasing, especially in lowland rice, and there is a pressing need to find a solution.

PSP-funded research exploits the discovery that plants have a natural antiviral defence system that is conceptually similar to the immune system of mammals, although the underlying mechanism is quite distinct. This antiviral defence can be triggered not only by viruses, but also by transgenes based on viral genes. The mechanism, known as homology-dependent resistance, does not require the expression of protein from the resistance transgene. It is therefore possible to produce RYMV resistant transgenic plants that do not express viral proteins. The approach results in a minimal and defined change to the rice genome with no detectable RNA (viral or transgene) or protein in the plants, thus greatly minimising conjectured hazards associated with their field release.

Clean-gene Technology

Half the world's population depends on rice as its major source of nutritional calories. Rice transformation technologies hold great promise for increasing rice productivity, especially in areas where rice farmers have little means to counter damage caused by pests and disease. The absence of classic plant breeding solutions (limited genetic sources of resistance available) and the limitations of chemical treatments (not economically feasible under low-input sustainable systems and extreme damage to the environment) present an excellent opportunity for biotechnological solutions.

PSP-funded research has been successful in finding a way of producing transgenic crops that are free of undesirable selectable marker genes (such as antibiotic resistance genes) and containing simple transgenic loci. This overcomes a constraint to the employment of genetic engineering: the perceived risks from introducing antibiotic resistance genes, which are only included because they are part of the transformation process, into the genetically modified crop.

Rice is the most extensively grown crop in South Asia, occupying nearly 50 million ha. Much of it is grown in the kharif (rainy) season. A substantial part of this area remains fallow during the rabi (post rainy) season because of several limitations, the prime one being limited availability of soil moisture. Precise estimates of such rice-fallows and their spatial distribution are not available. Since rice is grown on some of the most productive lands of this region, there is substantial scope to increase cropping intensity by introducing a second crop during the rabi season.

A review of existing technologies indicates that it is possible to productively cultivate legumes in most of these identified rice-fallows. An economic analysis has shown that growing legumes in rice-fallows is profitable for the farmers with a benefit-cost ratio exceeding 3.0 for many legumes. Also, utilizing rice-fallows for legume production could result in the generation of 584 million person-days employment for South Asia (66.7 in Bangladesh, 503 in India, 10.2 in Nepal, and 3.6 in Pakistan; values are in million person-days). Thus, introducing legumes into these rice-fallows will have a multi-faceted impact on the economy through employment generation, poverty alleviation, food security, quality of nutrition to humans and animals, and contribution to the sustainability of these production systems in South Asia.

Chapter 7

Techniques for Rapid Vegetative Multiplication

The main aim during pasture establishment is to provide an environment which is favourable for seed germination, seedling emergence and growth or that which is favourable for planted vegetative material to initiate new roots and shoots. Favourable conditions for establishment imply the absence of moisture stress and are often associated with good soil/seed contact induced by rolling and sufficient aeration. Establishment from seeds is usually more difficult in the tropics than under temperate conditions because seed is often not in a ready supply and they are mostly small where drought can kill small weak seedlings although this can also happen under the drier temperate climate.

Problems are also related to poor quality seed, seed dormancy or hard seededness or short life of the seed unless good storage facilities exist Many tropical grasses are commercially propagated by stem cuttings, rootstocks, rhizomes or stolons. This is primarily because early establishment is attained in tropical grasses when planted from vegetative parts. Some of the species that are propagated vegetatively include Digitaria decumbens, Brachiaria mutica, B. decumbens, Panicum maximum and bred varieties of Cynodon dactylon. Tropical legumes, unlike many tropical grasses are easily established from seeds.

Many are prolific seeders. However, there are legumes which are propagated vegetatively such as herbaceous legumes like Pueraria phaseoloides, Stylosanthes guianensis, Desmodium ovalifolium and tree-like legume, Leucaena leucocephala.

VEGETATIVE PLANTING MATERIALS

The choice of materials to be used will depend upon species, type of establishment being undertaken and availability of the planting materials. The nature of the growth form usually determines the response of a grass to cutting and management and may determine the type of planting material available.

Stolons

Stoloniferous grasses send out stems or stolons along soil surface, rooting at the nodes and producing new shoots. B. mutica, B. brizantha and I. aristatum are usually established from cuttings. Stems lying close to the ground should be pulled up and cut as these are likely to have existing root systems. Leafy material is not suitable as only the stems produce roots and start growing.

Rhizomes

Rhizomatous grasses produce stems or rhizomes below soil surface (e.g. guatemala grass, Tripsacum laxam). Rhizome propagated grasses grew rapidly in the first year; they had more vigorous buds, thicker stem and rhizomes and stronger tillering abilities than seed propagated grasses. Rhizomes can be obtained by digging them up in native pastures or in cultivated grassland. clearing and cutting them into 5-l0cm length with at least two nodes on each rhizome.

Tufts/rootstocks

Bunch of tufted grasses usually form clumps or tufts; if allowed to set seeds they will spread when seeds fall to the ground and germinate. The area from which the rootstocks are to be taken should be ploughed or rooted up and the clumps of the grass carted to the land on which they are to be planted. They are split into tufts of three to four tillers and planted by hand in the furrows. When the existing plant is divided to produce a number of pieces or root stocks (with root attached) the whole plants or the majority of the plant may be dug out of the ground. Pieces are mainly used for bunch grasses like guinea grass.

Hardwood/Softwood Cutting

The cueing from a woody stem of the previous seasons growth, from the branch or twig is an easy and inexpensive way of vegetative propagation. Leucaena sp. and Gliricidia sp. are usually propagated by this method.

PREPARATION OF PLANT NURSERY

When a large area is to be planted, the supply of planting material may pose a problem as it is not always easy to obtain rootstocks or cuttings sufficient for very large areas. Under this condition, the establishment of a plant nursery which serves as a multiplication area should be established between six months to one year ahead of the time at which the planting material is required. The ratio of land area devoted to nursery and land area to be planted varies according to the species and growth rate of the grass or legume itself as well as the planting system used. In a low density planting, in Western Samoa an establishment of one hectare of B. brizantha is required for 20 hectares of pasture to be sown. For high density planting one hectare nursery of B. brizantha could supply sufficient materials at each cutting to plant a further 15 to 20 hectares.

One hectare of B. mutica should be sufficient to plant 15 hectares of new ground. Williamson and Payne calculated that one hectare of guinea grass spaced at 50cm x 50cm should provide sufficient planting material for 24 to 30 hectares of land if each parent plant is split into four tufts. Na Sen and Lee Yan Qin estimated that to plant one acre of grassland, 3 sq.m of rhizomes is required whereas to plant an acre using seeds 18 to 30 kg of seed is required which need to be purchased at high price. Crowder and Chheda estimated that four men can collect in a single day sufficient vegetative materials for one hectare and six men can transplant one hectare in a day.

Normally a farmer obtains a small quantity of seeds or a few patches of vegetative material to establish a nursery then transplant the seedlings in 1.0m rows at about 0.50m within the row. After one or two growing seasons the plants may be dug and divided into crown splits for additional transplanting. This method is used on smallholdings where daily cutting is practised. The nursery should be given the best management so that within a year or two, this nucleus area would be able to provide material for further planting. It is advantageous to apply extra fertiliser to areas which will be supplying planting materials two to three months before planting time, to ensure good cuttings are available.

PLANTING TECHNIQUES

All methods of land preparation are best carried out as close as possible to the start of the wet season to take advantage of the long wet spell for the establishment phase. On arable land and level sites the land can be prepared to form a friable seedbed with furrows at 0.50 to 1.0m apart to place the propagules in the furrow and covered by hand or a cultivator. Normally the cane or stem cuttings such as those of napier may be planted by sticking them along furrows as in the planting of sugar canes. Planting along furrows will facilitate post-emergence cultivation to control weeds. Furrows are spaced about 75 cm apart and canes with two or three nodes are planted 30 to 50 cm apart within the row.

On steep slopes a furrow can be opened along the contour and the vegetative materials laid along the side of the furrows so that the tips extend above the soil when covered with a hoe or spade. Proper contact between the planting material and the soil is attained by disking the cutting under. The effects of partial (at least one node exposed) and complete covering of stolons and of compaction after planting of B. decumbens and B. humidicola. It can be seen that complete coverage of the stolons was very detrimental in both species. Compaction after covering has a large positive effect in the case of B. humidicola, but little effect on B. decumbens.

In another trial, one ploughing, and one rotovation to a depth of l5cm followed by planting 1-foot long stolons at lm x lm gave the best establishment in terms of establishment percentage, fresh weight yield and percentage cover. The method which had no tillage gave the poorest results in establishing Brachiaria sp. In the case of para grass, the planting materials can be broadcasted ova the cultivated land. The planting material must not be cut into small pieces. It is essential that disking be done as soon as the cuttings have been spread and since exposure to the sun ova a lengthy period will adversely affect establishment.

Anon suggested that disking controls weeds and favours formation of new tillers in stoloniferous species like pare grass. Good results of pare grass and Alabang X can be obtained when the cuttings are planted in fairly dry ground provided that they are harrowed at least 10 cm beneath the surface of the soil so that seedbed preparation has been very thorough and there is no weed competition. Planting machines used ranging from modified sugar cane, tobacco or cabbage planters to simple machines consisting of a tool bar with two or three tines and press wheels surmounted by a partly enclosed platform for carrying planting material.

The tractor proceeds at a walking pace, the planting material is put by hand into furrows formed by the tines and covered by soil pushed over the furrow by the feet of the planters walking behind the machine. Cuttings of some species like B. mutica and I. aristaturn can be scattered on bare soil and disced in by tractor and disc harrow. When legume pieces (e.g. hetero) are used, these could be inter-planted with the grass in alternate rows or is grown together in a nursery, planted simultaneously. The patches of a grass species like guinea grass may be expanded if they are lightly harrowed after the grass has seeded and after they have been oversown with the seed of a suitable legume. When this technique is used, grazing should be fairly light for a year or two, until the sward has thickened up.

Stem and stolons cut into sections with one or two nodes readily develop roots when placed in a proper medium with added nutrients. Parlridge reported that in Hawaii the technique has been developed using Jiffy ® pots of peat compound to allow more random broadcasting of cuttings without the labour input of digging. The pot can be spread by hand from horseback or even from the air. After spreading, trampling of pots by cattle into a wet soil may assist rooting. Dipping the pots in a nutrient solution before spreading or adding cuttings of a creeping legume like hetero to the grass slip is a good practice.

In plantations where the tree spacing is 10 metres, 3 runs with the subsoiler between each row of trees is recommended (the subsoiler can be run to within 1.5 metres of the trunk of the palm). Once ploughing, disking and subsoiling has been carried out, the land is usually regarded as being ready for the planting of the pasture species. Thorough cultivation will encourage a rapid and successful pasture establishment and also improve the thrift of the palms. Deep subsoiling will open up the soil and improve aeration and drainage; it will break up the mat of fibrous roots and encourage the development of new ones. Native or naturalised grazing land can also be burnt or heavily grazed after which vegetative pieces are placed in holes opened at random by hand or machines.

Depth of Planting

Cuttings and pieces are best planted at least 10-15 cm deep. Failure to plant the grass at a good depth may cause moisture stress as the soil dries out from the surface with resultant withering of plant material. Cuttings should be used on the same day or within 24 hours of cutting. Rootstocks, pieces and stakes can be kept for 2 or 3 days depending upon the species. Napier grass canes, for example, can be kept for some weeks under the right conditions. All planting materials should be kept in the shade before use.

When stem pieces with three to five nodes are set in the soil, it is usually the uppermost node that produces a bud because of apical dominance. This generally occurs in 8-14 days, to be followed by new shoots from the lower nodes. If the stem piece is rather juvenile, the upper portion may die back so that shoots form at the lower nodes. They may even come from nodes placed below ground level. Even though the upper nodes produce the first shoots, the lower ones usually grow rapidly, because of proximity to the soil water and nutrient supply. When long stem are layered in furrows and covered in soil, or left exposed on the soil surface, the f rst nodal buds develop near the stem base rather than the tip and new shoots do not develop from all nodes.

Size of Planting Material

With P. purpureum and T. laxam, stem cuttings or canes of 20-30cm in length with at least two or three nodes taken from mature stems are pushed into the soil at an angle. Root stocks or pieces can vary in size but must contain sufficient material to regenerate. The effect of planting piece size of tall guinea on early regrowth is shown in Table 1.

Although bigger pieces are vigorous and initially produce more regrowth, differences are less by second harvest. For easy machine handling pieces of 2.5cm diameter and 30cm length are chosen.

Table 1: Effect on early regrowth caused by size of tall guinea (P. maximum) planting pieces

Size of planting material (cm)		First harvest (kg)	Second harvest (kg)
Diameter	2.5	2.70a	2.71a
	5.0	4.15a	3.18a
	15.0	1.10b	1.73ab
	30.0	1.27b	1.87ab
Length	45.0	1.42ab	1.67a
	60.0	1.72a	2.13a

Legume may also be propagated vegetatively. Stylo (Stylosanthes guianensis) can be established vegetatively from stem cuttings. Tree legumes like L. leucocephala and G. macu!ata can be propagated from stakes or stem/branch cuttings which subsequently root and sprout. Chadhokar indicated that rooting of stem cuttings takes place about 6 weeks after planting.

Plant Spacing

Many different spacings are used for cuttings, pieces and stakes. Since bunch grasses like guinea grass do not send out stolons, pieces should be planted close together at about 1 m and later be allowed to set seed to cover the ground between. For rapid establishment of planted cuttings (either by hand or machine) a 1 m spacing is also recommended, although with species like B. mutica or I. aristatum broadcasting followed by discing may be used.

In general, the closer the spacing of cuttings or pieces the faster the establishment of a complete pasture. For bush legumes like leucaena, spacing will depend on the use. For grazing; double rows (1 m apart) between each two rows of coconuts, and for cut forage; closer spacing at 1 m between each row of leucaena are recommended. However, in order to ensure good legume establishment it is best to sow the legume before grass cuttings are established in order to reduce competition.

GRAFTING

Its principle use is to perpetuate varieties that are not successfully reproduced by cuttings, division, layering or producing no viable seeds. The process of grafting involves putting a piece of the desired plant (a scion) and the understock (parent plant) in unison. The idea is to fit two pieces of living plant tissue together so that they will grow and subsequently become one plant. All grafts are taped with plastic grafting tape, both for support and to stop possible water-borne fungal infection. Post-grafting care includes regular light watering and applications of soluble complete fertiliser. Tapes are removed by cutting vertically when cellulases are well developed. Generally, grafted plants are ready for transplanting into the field within three or four months of grafting.

Versace has developed the reliable techniques of grafting. This includes cleft grafts and whip and tongue grafts for leucaena. Seedlings about three months old (stem diameter 4-5 mm) make the best root stocks, while scions can be taken from similar seedlings of shoot of older established trees. Slow growing plants, particularly of species other than L. Ieucocephala often grow much better when grafted on to L. Ieucocephala root stocks. Grafting is also used to ensure the survival of valuable introductions and to multiply sterile or nonflowering genotypes.

GROWTH REGULATORS

Plant growth regulators fall into five classes which includes auxins, gibberellins, cytokinins, inhibitors and the gas ethylene. In most plants, the outstanding effect of the gibberellins is to elongate the primary stalk. This effect is caused by an increase in cell length, an increase in the rate of cell division, or a combination of both, depending on the specific types of plant treated. the grass plantis treated with gibberellic acid, tillering is reduced and the plant become more erect and stemmy. Herbage dry matter yields of some tropical grasses (Digitaria decumbens and Pennisetum clandestinum) treated with gibberellic acid were higher than the control when gibberellic acid was applied in the cool season.

Gibberellic acid treated plants were taller with longer internodes and modified leaf distribution. Spraying with gibberellic acid at 1000 ppm one week after sowing induced maximum height of the seedlings (15.7cm in 8 wks) compared with untreated seedlings (10.7cm). The difference in height was due to elongation of internodes. Alcantara suggested that L. leucocephala and D. virgatus which were difficult to establish from seed, could be propagated vegetatively in conjunction with growth stimulants; shoot growth of L. leucocephala and D. virgatum was stimulated by IAA and TP and by IBA and TP; respectively and root growth by IBA and TP and IBA and commercial IBA, respectively. Singh and Paliwal found that the lenticels of L. Ieucocephala attained the highest area in the main cuttings with gibberellic acid and in the side shoots with IAA.

Izham et al. found that cutting of Stylosanthes macrocarpa grew well in vermiculite as a rooting medium. However, the use of a synthetic hormone, which contained a fungicide, had no effect on rooting of S. macrocarpa.

MICROPROPAGATION

Micropropagation is the growing in vitro of minute/small (size lmm-3mm) propagules from shoot top, seedling tip, nodal segments, terminal to axillary buds with media containing nutrients normally required by normal plants. There are several factors to consider prior to the selection of the explants to ensure success. Some of the factors to be considered are availability, ease of sampling and handling, size of explant, ease of sterilisation, the development stage of the tissue, the physiological/nutritional and cytological state of the tissue.

Tissue culture work on leucaena has been initiated in laboratory with limited success. This is because rooting is a problem for leucaena and tissue culture is used to obtain the rooted shoot plants. Datta and Datta however, have developed a successful method of regeneration of L. leucocephala cv. Hawaiian Giant (K8) from the nodal explants. Healthy leafy shoots were obtained when BA or kinetin were added at lmg/ litre and healthy roots were obtained on medium with added IAA. The same species produced regenerants with vigorous rooted shoots when double the usual amount of KNO3 was added supplemented with lmg IAA/litre or2mg NAA/litre of medium solution. Venketeswaran et al stated that production of roots and shoots could be controlled by the manipulation of hormone ratios. They have developed techniques for producing as many as 3000 plants per seedling per year. Dhawan and Bhojwani successfully transferred the micropropagated plants of L. leucocephala to soil.

VEGETATIVE PROPAGATION

Distribution of material for vegetative propagation is seldom a commercial enterprise. On a local basis, a governmental agency or a livestock farmer will sell vegetative pieces of grasses usually in small quantities. Sometimes farmers are subsidised by government agencies in the purchase of planting materials or it could be obtained from research agencies. Stoloniferous grasses such as Pangola, Para and Kikuyu can be vegetatively propagated for commercial distribution. Grass propagules which is set into 'Jiffy' pots resume growth and are ready for transplanting within

3-4 weeks are easily handled for shipping, short-teml storage and direct transplanting without removal from the pots. In micropropagation, it is anticipated that the use of micropropagules will play an important role in extending the current application of plant tissue culture technology to gerrnplasm exchange between tropical and temperate countries and so lead to the more rapid introduction of new improved germplasm into the field and ultimately to its intended destination - the farmer.

CHAPTER 8

ENVIRONMENTAL ACCOUNTING FOR AGRICULTURE

Conventional agricultural accounts allow a macro level follow-up of the economic functions, but offer no insights in the multifunctional contribution of the agricultural sector to society, nor do they reflect the true costs and benefits of agricultural production for society. This chapter presents an integrated accounting system that provides a comprehensive perspective. A linkage is made between economic data from market transactions and both the costs of environmental bads and the benefits of environmental goods.

A general framework that uses quantitative internalisation of external effects is developed. The Economic and Environmental Accounts for Agriculture (EEAA) are built from the conventional agricultural accounts. Non-trade concerns, valued in monetary terms, are exogenously introduced. Relative shares can then be appreciated of those environmental goods and services that are already subject to market exchanges, those that are partially "monetarised" through public intervention and those for which there is no market or public transaction (and thus require proxies like avoidance costs).

The conventional Economic Accounts for Agriculture focus on the measurement of economic performance and growth as reflected in market activities and their evolution over time. From these accounts, indicators such as the gross value added (GVA) or agricultural income are calculated. Table 1 shows an example of the EAA for Belgium. The calculated indicators are often interpreted as the agricultural sector's contribution to society. However, the sector's contribution to the mere welfare is only one component of its significance to societal well-being. Human well-being is also influenced by other factors, such as the quality of the environment. The conventional accounts do not feature such factors and thus ignore a main part of agriculture's multifunctionality.

Table 1: The EAA for Belgium in 2001
(in million euro)

	Output value at basic prices (incl. subsidies and taxes on products)	7,376,33
—	Intermediate consumption b.p.	4,562,14
	Gross value added at basic prices	2,814,19
—	Consumption of fixed capital	606,18
	Net Value Added at basic prices	2,208,02
—	Compensations of employees	321,40
+	Other subsidies on production (grants for interest relief, for set-aside, agro-environmental subsidies, etc.)	127,45
—	Other taxes on production (taxes on property, vehicles, pollution, etc)	16,96
	Net operating surplus (factor income)	2 318,51

Even the few internalisations of environmental concerns that currently enter into the EAA are not treated univocally. Most environmental taxes are taxes on products and thus enter into the basic price of the intermediate consumption and into the

GVA. Most environmental subsidies are granted for specific methods of production and thus enter into the bulk of "other subsidies on production" and not into the GVA. Some environmentally inspired subsidies, such as the grant for cessation of animal production, are not entered into the EAA at all.

To provide policy-makers and public debate with a more comprehensive assessment of sustainable growth and development, first the environmental subsidies and taxes need to be emphasised, secondly the scope and coverage of economic accounting need to be broadened. In order to reflect agriculture's multiple functions, it becomes necessary to set off economic data (monetary values) against social and environmental data (physical values). To be able to measure the impacts and responses of a policy correctly, sociological and environmental information needs to be made consistent with the economic monitoring.

ECONOMIC ACCOUNTING FOR THE ENVIRONMENTAL OUTPUTS

The main problem in balancing all costs and benefits is the conversion of various physical data to the same units as the economic accounts. Up to now several attempts have been made to assign monetary values to the negative externalities of agriculture, e.g. pesticide use; nutrient leaching or erosion. Environmental economists have also developed methods for valuing environmental service functions, such as willingness-to-pay techniques. Unfortunately, these have rarely been applied in comprehensive studies evaluating the aggregate positive side effects of agriculture.

In Belgium, Vanslembrouck, gives an economic assessment of landscape amenitics provided by farmers. Difficulties in valuing the positive externalities of agriculture are undoubtedly the reason why they are often under-emphasized in economic assessments. Until recently the environmental costs and benefits of agriculture had not been integrated into a comprehensive framework linking them with monetary dimensions at macro or sector-wide level. In contrast, some whole-economy attempts to integrate economic and environmental information were made, such as SERIEE. However these systems lack sufficient detail to deal with the different agricultural productions and they are limited to those environmental aspects that can directly be expressed in monetary terms.

Both accounting systems are based on the whole-economy System for Environmental and Economic Accounts. This framework complements and expands the System of National Accounts to include environmental assets, non-commodity assets are monetised and both physical and monetary measures are integrated. The eftec study provides monetary estimates of both the positive environmental services provided by agriculture and the negative flows resulting from over-use of natural assets. Distinction is made between impacts on other sectors and impacts on society's welfare in general. The first, are valued trough market data, like purification costs, damage to roads from soil erosion, damage from flooding, etc. The latter are valued through non-market data derived from household willingness to pay (WTP). However, no effort is made to integrate these values into the economic accounts.

This framework puts agricultural production in the centre and links the environmental outputs as closely as possible to the agricultural activities. This has the advantage that the accounts might be split up in different sub-sectors, by which means the activities where alterations need to be made can be identified more easily. In or opinion, the mean weakness of the eftec approach is the lack of articulation between the classical economic accounts and the monetary estimations of environmental impacts. Because the income accounts are kept as a "black box" and the monetised externalities as satellite accounts, interpretation possibilities for the results are rather limited.

Multifunctional Accounting

The accounting framework starts from the conventional Economic Accounts for Agriculture. The non-commodity outputs are internalised into the EAA, which thus becomes extended to Economic and Environmental Accounts (EEAA). These "multifunctional" accounts are fully expressed in monetary terms and integrate on the one hand the costs of negative and on the other hand the value of positive externalities with the value added of agricultural production. The EEAA produces a "multifunctional value added", enabling a complete evaluation of sustainability over time. This study goes beyond most of the "green accounting" efforts, as it internalises not only the negative environmental effects, but also the service or amenity function of agriculture. Building the EEAA involves three main research topics:

1. *Definition of the system boundaries*: not all externalities of agriculture activities need to be taken into account, either while their links with the activities are rather vague or too indirect (e.g. the depletion of natural resources for tractor manufacturing) or while estimation difficulties outweigh relevance (e.g. consumer's avoidance costs in relation with pollution, like the choice to drink only bottled water).
2. Association of flows of identified environmental goods and services and flows of pollutant emissions with monetary flows. Monetary values are derived from proxies like subsidies, taxes, treatment and restoration costs, etc. and from indirect valuation ("willingness to pay" approaches).
3. Integration these values with direct (market) flows into a single framework and calculation of aggregate indicators, such as a "multifunctional GVA".

From the EEAA, two types of analyses can be implemented. First, it becomes possible to appreciate the overall impact of multifunctional aspects of agriculture at national or regional level. This allows simulating scenarios (e.g. increasing agri-environmental subsidies), taking into account either the impact on agricultural income (through the effect on production level and the premium itself) or on societal welfare. Secondly, it allows evaluating the share of different actors (producers, public authorities, water agency, consumers, etc.) in the costs or benefits of environmental goods and services provided by agriculture.

For example, the proposed framework emphasizes some aspects that are already internalised in the conventional accounts, like agroenvironmental subsidies and taxes. In the conventional EAA, agriculture's income increases when subsidies increase. However, these subsidies are costs for society. In the EEAA the two effects are linked and can be compared. Thus, this approach allows for a better way to deal with financial transfers from society to farmers and conversely.

While building the EEAA framework, the environment is considered to provide three functions to agriculture and to the economy and mankind in general: source, sink and service functions. Furthermore the agricultural system is evaluated according to the driving force - pressure - state - impact and response or DPSIR cause-effect chain. The P- and S-indicators are typically measured in physical values. The I-indicators might be valued in monetary terms, through avoidance or restoration costs for the source and service functions or through revealed or stated preference techniques (WTP) for the service functions.

The R-indicators, the responses from the government, often consist in subsidies paid to (or taxes imposed on) farmers for adopting (or abandoning) certain (less) sustainable practices. Some, but not all of them already enter into the EAA. The general framework, with the different accounts proposed, is represented in figure 1.

The EEAA are built in three stages.

The first stage comprises the monetary and physical accounts. The starting point is the conventional EAA, with the monetary driving factors (GVA etc.) and already including some subsidies and taxes (table 1). From the EAA and external information on subsidies and taxes that are not yet internalised, the Environmental Protection Expenditures Account (EPEA) is derived. It reveals all the expenditures and investments concerning the environment, especially the public ones. Next to the monetary accounts, satellite physical accounts are drawn up. They contain information on both pressure and state indicators and are differentiated into 3 accounts in order to reflect the 3 environmental functions.

For example, the Pollutants and Residuals Account, reflecting the sink function, contains things like the nutrient surplus on the soil balance and the nitrate concentration in surface water. The Asset Account, reflecting the service function, contains for example the number of endangered animals/plants being kept/cultivated on farms.

In the second stage, monetary values are assigned to the positive and negative externalities in the physical accounts. Previously elaborated valuation methods are applied to the non-tradable by-products of agriculture. Together with the EPEA, these monetary values form the Non-Tradable Output Account (NTOA), which represents the economic value, i.e. the costs and benefits of the environmental goods and bads. The eftec study stops at this point.

In the third and final stage, the NTOA is combined with the conventional EAA. At that point the complete DPSIR cause-effect chain is integrated into a single accounting system and the Economic and Environments Accounts for Agriculture (EEAA) arise. From these accounts indicators allowing a more "social" policy evaluation, such as the "multifunctional value added", can be calculated. In contrast to the current EAA, which

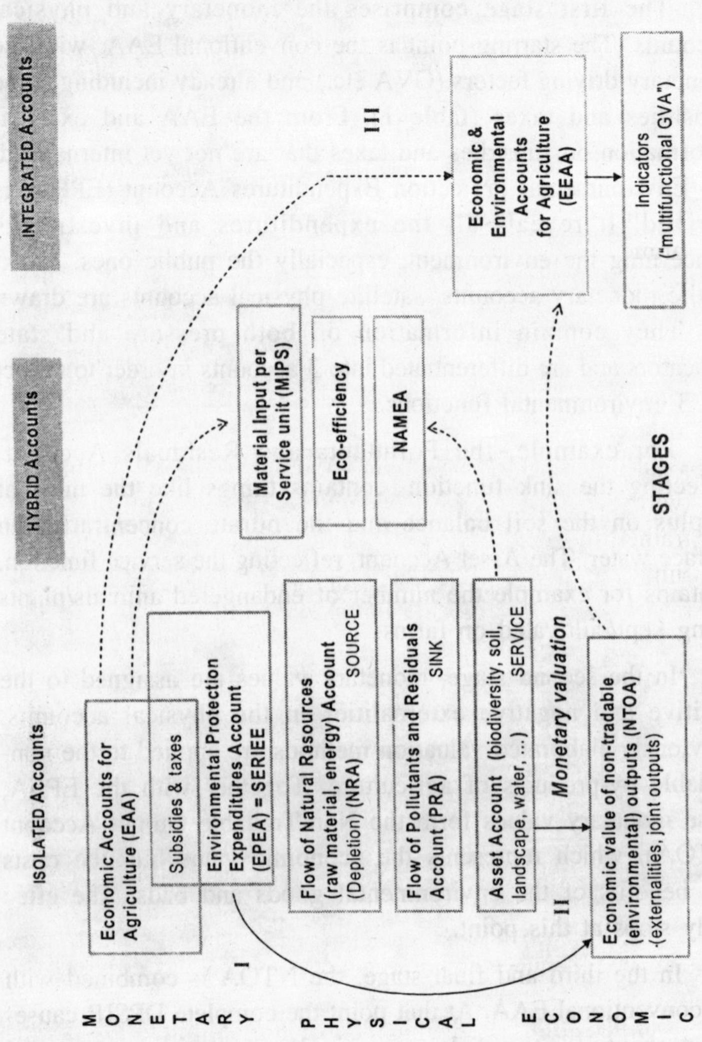

Figure 1: General framework of the Economic and Environmental Accounts for Agriculture (EEAA)

only records private costs and benefits, the expanded EEAA also include the external environmental costs and benefits, i.e. the cost and benefits the rest of society. Table 2 gives an example for pesticide use (with negative externalities) and for landscape amenities (positive externalities).

Agro-environmental Subsidies and Taxes

In the eftec framework, they are not included. Atkinson et al. argue that the value of environmental impacts should be recorded regardless of the fact that they may be "internalised" through subsidies or taxes. Their primary rationale is to account for changes in natural assets, regardless of the policy measures in place, since it is not currently clear if taxes "over"- or "under"-regulate environmental impacts. Taxes can not be assumed "optimal" in the sense of internalising an externality consistent with the economic optimum; on the contrary, they represent a rather arbitrary form of "internalisation". However, even if it is unlikely that subsidies and taxes reflect society's true WTP, they are an important factor affecting the amount of an environmental function that is produced by the agricultural sector. When for instance nutrient surpluses are strictly taxed, the sector will try to abate its nutrient production or when small landscape elements are heavily subsidised, this is an incentive towards providing this service to society.

So there certainly is a linkage (even if very difficult to establish) between damage costs or WTP for a better environment and subsides and taxes. When the community accepts to pay farmers for less intensive agricultural practices, the damage and restoration costs for society will probably decrease, but the avoidance costs (here, contribution to the agricultural policy budget) increases. Government's WTP should be considered as much as private (household) WTP. Precisely because it is not clear if any tax or subsidy internalises environmental impacts in a economic optimal way, it is very useful and important to highlight these transfers between farmers and rest of the society and to record them together with other costs and benefits (table 2), when comparing environmental agricultural accounts in different spaces or time.

From an accounting point of view, subsidies increase farmer's Net Value Added and decrease the financial means of public authorities for preventing or repairing environmental degradation (and consequently increase taxation of citizens). In the EEAA, environmental subsidies and taxes appear only as transfers from one part (External social costs) to the other (Private costs and benefits). Therefore the agroenvironmental subsidies and taxes are recorded twice, once with a positive and once with a negative sign.

Table 2: Examples of the EEAA for pesticide use and landscape amenities.

Pesticide use		Landscape amenities	
Private costs and benefits of the agricultural sector (EAA)			
	Output value of plant production		Total output value
-	Intermediate pesticide consumption, including tax on some pesticides	-	Total Intermediate consumption
	Gross Value Added at basic prices		Gross Value Added at basic prices
-	Consumption of fixed capital	-	Consumption of fixed capital
Net Value Added at basic prices			Net Value Added at basic prices
+	Environmental subsidies: organic farming, integrated fruit production, mechanical weed killing, etc.	+	Environmental subsidies: Installation and/or maintenance of small landscape elements
-	Environmental taxes: contribution for recycling recipients	+	(other subsidies – other taxes) on production

+	(other subsidies – other taxes) on production		
–	**External social cost**	+	**External social benefits**
-	Environmental subsidies	-	subsidies
+	Environmental taxes		
Avoidance cost	cost of reducing production or pesticide use	+	Value attached to landscape by society:
Abatement cost	cost of reducing environmental impact, e.g. by new products or new technology		value of living in the countryside, value of surroundings for rural tourism,
Treatment cost	end-of-pipe costs of decreasing the discharge of pollutants in the environment, e.g. cost of retrieving recipients		recreational value,etc. —Mostly non-market values — valuation trough
Restoration costs	expenditures by third parties for restoring degraded natural systems cost for drinking water purification		"willingness to pay" approaches: hedonic price analysis
Damage costs	costs linked to over-use of environmental sinks, e.g. biodiversity loss, health problems		travel cost methods etc.
=	Multifunctional net value added		

The principal interest of the EEAA is not as much the absolute value of each section, but their relative weight and the overall evolution of their share over time. Therefore agro-environmental subsidies and taxes explicitly appear, despite their neutral effect on the total. One of the main purposes of the EEAA is to be able to monitor in long run which sections' importance is decreasing and which one is increasing.

The physical accounts are drawn up as in classical environmental reporting. The next step is to elaborate the Environmental Protection Expenditures Account (EPEA). This elucidates the transfers between the government and the agricultural sector and overcomes inconsistencies in the conventional accounts mentioned above. The EPEA contains all environmental subsidies and taxes, irrespective of whether they already are in the EEA. For Belgium in 2001, the total environmental subsidies amount to 59 million euro, the total taxes to 19 million euro. Thus the net government expenditures amount to 40 million euro, i.e. 1.8 % of the net value added of the agricultural sector. Those environmental effects where market transactions are involved. But for those externalities, for which there is no market, people's willingness to pay (WTP) for amenities or to avoid damage needs to be assessed.

Chapter 9

Role of Agricultural Modernisation

The importance of agricultural sector is not calculated in a traditional way by measuring units of agricultural output per unit of labor or land based on some index of inputs. One reason for this is such measures of agricultural productivity are not available for a number of countries. Thus a different and unique perspective on agricultural productivity is introduced here. Here the importance of agriculture is measured by the amount of investment made on land to modernize it and thereby make it productive.

Theoretically this may be more appropriate than the traditional measures. This is due to the fact that agricultural productivity could be high because a particular nation is well endowed with natural resources. High productivity, in this situation, will likely decline with depletion of natural resources unless some steps are taken to maintain this productivity.

On the other hand, it is being hypothesized here that countries which made some effort towards improving or maintaining their agricultural productivity by investing in it and modernizing the agricultural sector would most likely be the ones who would continue to reap the benefits from increased agricultural productivity.

Much of the early thinking on economic development ignored agriculture altogether. The piecemeal establishment of manufacturing in poor regions lacking infrastructure would not likely be successful. Instead, investment in industry and manufacturing had to be on a broad front such that various industries could create markets for each other's products. There were a set of theories which, however, did see a role in the development process for agriculture. The modern sector was driven by profit maximization and the accumulation of physical capital.

The traditional sector was subsistence oriented and usually thought to be dominated by peasant agricultural production. This sector was characterized by output sharing mechanisms rather than profit maximization. In many of these models it was presumed that the traditional sector was characterized by surplus labor. There was so much labor in this sector that it could be withdrawn and put to productive work in the modern sector without any fall in output in the traditional sector. In effect, "free growth" was possible through mobilization of labor for modern production.

However, once surplus labor was exhausted, then the expansion of the modern sector might very well be strangled. Continued withdrawal of labor would lead to falling output in the traditional sector leading to a rise in the relative price of the traditional sector output relative to that in the modern sector. If the traditional sector produces mainly food, the rising relative cost of food would push up wages to the modern sector, cutting into profits, reducing investment and the expansion of this sector.

Johnston and Mellor built upon these ideas in their analysis of the role of agriculture in overall economic development. They argued that agriculture supplied the labor necessary to man the modern sector firms as well as the food necessary to feed that labor. In addition, the agricultural sector was seen as serving as a market for the produce of the modern sector, a stimulus from the demand side. Finally, perhaps most importantly, agriculture was likely to serve as the main source of savings necessary to finance the expansion of the modern sector. After these developments, agriculture disappeared from general models aimed at analyzing economic growth and development.

Instead, much of the literature concerned with agriculture concentrated on analyzing productivity growth in the traditional, agricultural sector. Perhaps the most interesting and innovative work in this area has been undertaken by Hayami and Ruttan in developing a theory of induced technical and institutional innovation. Recently, multiple sector growth models have begun to be constructed with agricultural sectors. Matsuyama developed an endogenous, two sector growth model. In this model the engine of growth, the driving force, was learning by doing in the manufacturing sector. He compared and contrasted the implications of a closed and open economy model.

In the closed economy case, an increase in agricultural productivity spurs overall economic growth since this eases the expansion of learning by doing via manufacturing. However, in the open economy case there is a negative link between agricultural productivity and overall growth. This occurs because the more productive the agricultural sector is, the more resources that are devoted to agriculture based on comparative advantage. This, of course, implies less manufacturing, less learning by doing, and less growth.

The results from Matsuyama's model are of course based on assuming that all learning by doing occurs in manufacturing, none in agriculture. However, learning by doing in manufacturing could enhance productivity in agriculture and perhaps vice versa. More generally, the model's results stem from the assumption that agriculture is, by nature, incapable of sustaining rapid productivity growth. Thus it is inevitable that higher initial productivity in agriculture would lower long-run growth.

This idea that productivity growth is slow in agriculture is actually contradicted by empirical analysis. In addition, "there is strong evidence of convergence in levels and growth rates of

TFP in agriculture, suggesting relatively rapid international dissemination of innovation". These results suggest that a large agricultural sector need not be a disadvantage in the overall growth process. It may likely be an advantage if productivity growth is rapid. Thus contrary to the assumption made by Matsuyama, the agricultural sector has significant prospects for rapid productivity growth.

Theorists have now begun to explicitly model the agricultural sector in multiple sector growth models. A recent example of this is provided by the work of Gollin, Parente, and Rogerson. They extend the neoclassical model so as to incorporate an agricultural sector. They attempt to model the structural transformation that comes with development (agriculture shrinking, manufacturing expanding). Agricultural output per person must reach a certain level before modern technology will be applied to agricultural production and labor can flow out of agriculture and into industry.

The rate at which labor can then flow out is determined by the rate of technological change in agriculture. Low agricultural productivity can thus substantially delay the onset of industrialization. Other research has examined new links between agriculture and the growth of the rest of the economy. One can think of these new links as representing non-traditional roles for agriculture. Timmer argues that agriculture plays a significant role in reducing poverty. The bulk of the poor reside in rural areas so an increase in growth in agriculture has a significant potential for reducing such poverty.

In addition, agricultural growth stimulates the development of agribusiness activities as well as stimulating the demand for manufactured inputs. Stringer further argues that the agricultural sector performs important social welfare functions in developing nations. For example, during an economic downturn or an external income shock or financial crisis, agriculture can act "as a buffer, safety net, and as an economic stabilizer". The flexibility of the production process allows for labor to be substituted for capital thus cushioning economic blows. Thus people frequently return to the farm during bad times.

First, does absolute economic convergence occur and, if not, is convergence conditional upon agricultural productivity. In other words, is increased agricultural productivity a condition for economic convergence. Second, if so, is the modernization of agriculture an important determinant of overall growth. Is this effect robust to the inclusion of other variables. Third, given the non-traditional roles for agriculture, does agricultural modernization have a significant impact on human development. Is this impact robust to the inclusion of other variables.

Test for absolute convergence, the following equation is estimated:

$$GR6095 = a + b\,(LnGDP60) + \varepsilon, \qquad (1)$$

where GR6095 is the average growth rate of our sample countries for the time period 1960 to 1995. The right-hand side variable, LnGDP60, is the natural log of GDP per capita in 1960 for each of our sample countries. Of course, å is the error term. As is common in the literature, the sign on the right-hand side variable would tell us something about convergence. If the sign is negative and statistically significant, this would imply that those countries which have the higher GDP per capita will tend to grow slower than those countries with lower GDP per capita. Thus absolute convergence is taking place.

Most studies have found very little evidence in support of absolute convergence. However, there is a substantial literature that conditional convergence does occur. That is, once one accounts for specific variables that influence the long-run, steady state equilibrium, convergence may still be found (convergence to different equilibria). Another way to think of this is that there are certain preconditions that must occur before convergence takes place. It is hypothesized that productive agricultural sectors are a necessary precondition for economic convergence. In order to test this hypothesis, the following equation is estimated

$$GR6095 = a + b(\ln GDP60) + c(\ln Fert65) + \varepsilon, \qquad (2)$$

where lnFert65 is the natural log of fertilizer intensity for the year 1965. The latter is measured as kilograms per hectare of land. This variable is used as a proxy measure for agricultural productivity or the degree of modernization of agriculture in 1965. This proxy variable is used because direct data on agricultural productivity for a large number of countries for that time period is not available.

The intuition behind equation two is fairly obvious. It is hypothesized that high agricultural productivity or a modern agricultural sector is a necessary precondition for convergence to occur. Thus, it is hypothesized that the coefficient c will be positive and statistically significant and the coefficient for b will be negative and statistically significant. Another way of testing the same hypothesis is to divide the sample countries into two groups, those with above average and those with below average fertilizer intensity in 1965.

One would hypothesize absolute convergence for those countries above the average for fertilizer intensity and divergence or no trend for those with below average fertilizer intensity. Thus equation one will be estimated for the two groups. It is hypothesized that b will be negative and statistically significant for those with above average fertilizer intensity and b will be statistically insignificant for those below the average fertilizer intensity. If indeed agriculture appears to be a precondition for convergence, then the next step is to determine how robust is agriculture's influence on growth. This is tested by estimating the following equation by adding one variable at a time.

$$GR6095 = a + b(\ln GDP60) + c(\ln Fert65) + d(School) + e(\ln Inv6095) + f(Open\ 6095) + g(ICRGE80) + h(Statehist) + I(Asia) + j(LatA) + k(SSA) + \varepsilon. \qquad (3)$$

The additional variables that are added are a measure for educational attainment given by School which is the average years of total schooling of the population from 1960 to 1985, a measure of the investment given by lnInv6095 which is the natural log of average investment from 1960 to 1995, a measure of openness of the economy given by Open6095 which is the average ratio of exports plus imports divided by GDP for the period 1960 to 1995, a measure of institutional quality given by ICRGE80 which is an the average of five different measures of institutional quality, a variable for state antiquity given by Statehist, and a series of dummy variables: Asia (Asia), Latin America (LatA), and Sub-Saharan Africa (SSA).

Several of these variables need further explanation. The Statehist variable is an index of how long a nation state has been in existence for various regions of the world. The time period covered is from 1 to 1950 C.E. The higher the index number, the longer a state has been in existence. The ICRGE80 variable is an average of measures of corruption, repudiation of contracts, expropriation risk, rule of law, and bureaucratic quality for the year 1980. The higher this average, the better the quality of institutions. Much of the data used to estimate this and the previous equation are taken from Bockstette, Chanda, and Putterman.

Since it may be argued that the intensity of fertilizer usage may be a narrow measure of agricultural modernization, a more sophisticated measure of agricultural productivity or agricultural modernization was derived by gathering data from the World Resources Institute (who in turn gathered the data from the FAO) on tractor intensity, which is tractors per hectare and by including a measure of average years of schooling of the population in 1960.

The reason behind including a measure of education along with mechanization in farming is to recognize that increased human capital can result in better technology and a more efficient utilization of the available agricultural technology. These two additional terms are multiplied by the fertilizer intensity term to create an interaction term. It is expected that this interaction term gives a more accurate representation of the degree to which the agricultural sector was modernized. This interaction term is referred to henceforth as AgModern.

It was substituted into equation (3) for fertilizer intensity and the model was re-estimated to see if the same relationships are found to hold with a more sophisticated treatment of agricultural modernization. Since AgModern includes a measure of human capital, the School variables and Statehist variable are excluded from the estimations to avoid multicollienarity issues associated with these variables and the dependent variable.

Under such assumption one would expect that the level of human development would also be influenced by agricultural modernization. In order to test this proposition, per capita GDP growth is replaced with the growth of the human development index (HDI) from 1975 to 1995 as the dependent variable. The data used to calculate this measure is taken from various issues of the Human Development Report.

Equation (3) is re-estimated once again, but with the growth in the average human development index (HDI7595) as the dependent variable and with AgModern as a measure of agricultural modernization. Thus the following equation is estimated to test the hypothesis.

$$\text{HDI7595} = a + k(\text{HDI75}) + b(\text{AgModern}) + g(\ln\text{Inv6095}) + d(\text{ICRGE80}) + z(\text{Open6095}) + q(\text{Asia}) + l(\text{LatA}) + n(\text{SSA}) + \varepsilon. \qquad (4)$$

Notice that lnGDP60 does appear in the equation. In equation (4), HDI75 replaces lnGDP60 to capture the impact of initial HDI. The data set covers 89 countries. For various regressions some countries are missing values, thus the size of the sample accordingly shrinks.

In theorizing about long-run growth and development agriculture seemed to disappear from the literature and ceased to play an important role after the development of dualistic models in the 1950s and 1960s. However, recently there has been a reemergence of interest and a reconsideration of the importance of agriculture in the development process with a number of models being constructed which incorporate agriculture.

These models and the theories they propose find agriculture to play a critical role in long-run growth and development. Here analyzes the importance of a modernized agricultural sector to economic growth and development. Agricultural modernization is defined in a simple as well as a complex manner. In its simple form it is measured by the intensity of fertilizer usage per hectare of land.

In its more complex measure it includes improved technology which is a product of a measure of fertilizer intensity as well as tractor intensity along with a more educated population. The idea is that increased technology in agriculture would increase agricultural productivity or value-added in agriculture while freeing up resources which could be used for other productive uses.

The reason for including a measure of education is that an educated population adds to the overall human capital which thereby adds to the proportion of human capital that can be allocated towards achieving further technological progress in agriculture. The results indicate that the modernization of the agricultural sector is not only a significant precondition to achieving economic growth, but that agricultural modernization itself has had a significant positive impact on economic growth as well as in improving economic well-being. The implication then is that policy aimed at achieving higher growth rates should be aimed at improving agricultural productivity via investment in that sector.

Chapter 10

Agricultural Biotechnology in the Asia Pacific Region

The Asia-Pacific region has over 56 percent of the global population, about 73 percent of the world's farming households, but only 31 percent of the world's arable land. Per caput arable land availability in the region is only 0.26 ha against 1.51 ha in the rest of the world. The region is, however, more favourably endowed with irrigation as compared with the rest of the world. It accounts for about 61 percent of the world's total irrigated area; almost one-third of the total arable land is irrigated, as against one-tenth in the rest of the world (Table 1).

Agricultural production in the region moved at a much faster rate than in other parts of the world. It accounted for 44.4, 40.6, 28.9 and 53.4 percent of the world's productions of cereals, roots and tubers, fruits and vegetables, respectively. From 1982-92, production of these commodities increased at a much faster rate than in other regions; cereal production growth rate for the region was 2.3 percent as against 1.2 percent in the rest of the world. As regards meat and milk, the region's share was respectively 31.6 and 22.6 percent of the world's production, but the growth rates were 5.8 and 4.5 percent against 1.7 and 0.1 percent in the rest of the world.

The region accounted for 46 percent of the world's fisheries production; not only did it have almost a monopoly in the world's aquaculture production, 83.1 percent, but it also recorded a high growth rate of 8.6 percent against 5.1 percent in the rest of the world (Table 1).

The higher production growth rates were stimulated through the development and adoption of improved technologies, coupled with appropriate government policies and programmes. Average cereal yield in the region was 2 929 kg/ha against 2 689 kg/ha in the rest of the world. In fact, the growth in cereal production in the region during the past decade accrued essentially through increase in yield, while the area remained stagnant or even declined. This trend needs to be maintained in the future since there is little scope for horizontal expansion of arable land.

Despite the impressive growth in food and agricultural production during the past 25 years in the region, 523 million people, about one-fifth of the region's population, are chronically malnourished (FAO, 1993b). The concentration of malnourished people is considerably higher in South Asia than in East Asia (Table 2). The population of developing Asia is expected to grow to 3 726 million by the year 2010, comprising about two-thirds of the developing world. Although the growth rate will decline, the annual increment will still be more than 50 million people.

Projected production and demand in South and East Asia will be neck and neck, but demand will marginally outstrip production. This gap will be reflected in the number of chronically malnourished people in the different regions. By the year 2010, the number of malnourished in East Asia will drop to 70 million and in South Asia to 202 million from 252 million and 271 million in 1988/90, respectively (FAO, 1993b). In other words, while the number of malnourished in developing Asia will decline considerably, the region will still contain about 43 percent of the world's malnourished people.

Table 1: Selected indicators of agricultural development in the Asia-Pacific region

Indicators	Asia-Pacific	Rest of the world	World	Asia Pacific as % of the World
Total population (million), 1991	3,034.7	2,354.5	5,389.2	56.3
Agricultural population (million), 1991	1,753.8	656.5	2,410.3	72.8
Agricultural land (million ha), 1991	453.3	988.3	1,441.6	31.4
Ratio of agric. land to agric. popul. (ha/caput)	0.26	1.51	0.60	43.3
Irrigated land (million ha), 1991	146.5	95.0	241.5	60.7
Irrigated land as % of agric. land, 1991	32.3	9.6	16.8	-
Mineral fertilizer use (NPK kg/ha), 1991	126.6	77.5	92.9	136.3
Agric. prod. Index (1979/81 = 100), 1991	148.17	114.8	6 125.26	-
Cereal prod. (million tonne), 1992	866.8	1 085.4	1 952.2	44.4
Cereal yield, (kg/ha) 1992	2 929	2 689	2 791	-
Cereal yield growth rate (%), 1982-92	2.3	1.2	1.6	-
Roots and tubers prod.				

(million tonne), 1992	238.0	348.1	586.1	40.6
Roots and tubers yield (kg/ha), 1992	14 318	11 148	12 249	-
Fruit production (million tonne), 1992	106.9	262.6	369.5	28.9
Fruit production growth rate (%), 1982-92	3.6	0.8	1.6	-
Vegetable production (million tonne), 1992	243.4	212.7	456.1	53.4
Vegetable production growth rate (%), 1982-92	2.9	1.1	2.0	-
Livestock prod. index (1979/ 81=100), 1991	160	110	126	-
Meat production (million tonne), 1992	57.6	124.5	182.1	31.6
Meat production growth rate (%), 1982-92	5.8	1.7	2.8	-
Milk production (million tonne), 1992	117.5	401.6	519.1	22.6
Milk production growth rate (%), 1982-92	4.5	0.1	0.9	-
Fisheries production (million tonne), 1991	44.6	52.3	96.9	46.03
Aquaculture production				

(million tonne), 1991	13.8	2.8	16.6	83.1
Aquaculture product growth rate (%), 1981-91	8.6	5.1	7.9	-

Table 2: Estimates of chronic undernutrition in developing countries

Region	Year	Per caput food supplies (cal/day)	Total population (million)	Undernourished	
				% of total	Mill.
East Asia	1969/71	2 020	1 120	44	497
	1979/81	2 340	1 358	26	359
	1988/90	2 600	1 558	16	252
South Asia	1969/71	2 040	738	34	254
	1979/81	2 100	926	31	285
	1988/90	2 220	1 144	24	271
93 developing counts.1/1969/712		120	2 585	36	941
	1979/81	2 320	3 232	26	843
	1988/90	2 470	3 905	20	781

While the need for further intensification of agricultural production in the region is cleaı, the ways of the intensification process must change in order to avoid the adverse effects associated with the past. Yield increases in the past three decades were triggered by widespread use of high-yielding modern varieties, expanded irrigation and increased use of chemical fertilizers. The Asian farmer on an average uses 127 kg NPK/ ha against 78 kg/ha used in the rest of the world. These enhanced uses of irrigation and mineral fertilizers, which were often not really efficient, had their own negative side-effects such as soil salinity and nutrient leaching. With intensification, the biotic stresses of pests and diseases had also increased. Furthermore, about 70 percent of the total cultivated land is rainfed and subject to monsoonal vagaries and other abiotic stresses, and was generally bypassed by the "green revolution", thus exacerbating inequity.

In order to meet the unprecedented demand for food arising mostly from the burgeoning population and to meet development demands, forests were razed indiscriminately, causing a host of environmental and soil degradation problems. In Southeast Asia alone, 5,000 ha of tropical forests are cut every day. The region encompasses megacentres of biodiversity - the raw materials for biotechnology. But, because of wide adoption of a few often interrelated modern varieties, and because of deforestation, the treasure of biodiversity has been eroding fast. Because of soil degradation, increasing genetic vulnerability and intensifying pest incidences, there are increasing examples of plateuing off or even decline of yields, productivity a and profitability of major agricultural systems, such as rice-wheat cropping systems in several Asian countries.

Keeping the above developments and challenges in mind, it is clear that while the process of intensification of food and agricultural production in the Asia-Pacific Region must continue in order to feed its people, the ways of enhancing production must be altered. Encroachment on marginal lands and fragile ecosystems, deforestation, erosion of biodiversity and environmental deterioration must not only be avoided, but also reversed, in order to attain enhanced and sustained agricultural production. The "green revolution", with its strong positive and some negative impacts, was responsible for accelerating food and agricultural production, especially cereal production, over the past 30 years or so. But the revolution is now on the wane. At this juncture, biotechnology is considered by many to be the means of triggering the next green revolution.

In the context of intensifying the use of biotechnologies in Asian agriculture, Swaminathan has listed several features that confer comparative advantages to the region. First and foremost is the richness of biological and ecosystem diversity. The large population, especially women who could handle tissue culture and other simple biotechniques, provides a huge and not-so-costly labour force. Furthermore, there is a greater opportunity for "production-by-masses" rather than merely mass production to bring about equity. Finally, the most important industrial feedstock that most Asian countries have in rural areas is the agricultural biomass of plant, animal, fish or tree origin, providing vast opportunities for preparing value-added products.

In the Asia-Pacific region, future increases in agricultural production must accrue essentially through increases in yields. Yields can be increased by (i) preventing the pre-and post-harvest losses; (ii) raising actual yields closer to the current production potential; and (iii) raising the production potential. Biotechnologies are already being applied to one or the other, or all three options for yield increases in several countries of the region. Particular mention may be made of in vitro culture techniques in potatoes and cassava and plantation crops, haploids in rice, diagnostic kits for disease identification, new and recombinant vaccines, embryo transfer, increased productivity of fishes through sex reversal, polyploidy, hormonal treatments and disease control.

These techniques should further be refined/standardized and rendered more cost-effective to improve their transfer to and adoption by the majority of small farmers. Plant tissue culture techniques are particularly suitable for Asian settings. Transgenics for new yields, adaptability and quality, are being designed not only in the developed countries, but also in several developing countries in the region, which may be commercialized in the future. The first contributions of biotechnology towards higher yield will come by protecting plants from diseases and pests, thereby cutting losses.

Australia is contemplating the release of a pest-resistant transgenic cotton variety by the end of 1996. However, for commercialization of the products under development, resolution and/or establishment of regulatory aspects, particularly biosafety and intellectual property rights should be undertaken simultaneously. The important role of the private sector and the proprietary nature of many of the new products and processes are significant in the context of biotechnology research and development. These features have implications for developing countries using and developing biotechnology. In order to promote equity, mechanisms should be developed and adopted for recognizing and rewarding both formal and informal innovations.

Of the major crops of the region, rice is likely to benefit most through the use of biotechnology. This is happy news since rice is basic to food security in the region. Haploid production and embryo rescue techniques are already being exploited widely. Molecular marker-aided selection, DNA fingerprinting for identification of genetic variation in pests, pathogens and rice populations, protoplast transformation and production of transgenics for introduction of novel genes are in progress in the region.

Genetically engineered tungro-resistant rice lines are being engineered at IRRI. Other expected early successes are transgenic hybrids of Brassica and sunflower, and virus-resistant potatoes and soybeans. Other distinct possibilities are biotechnologically derived biopesticides, biocontrol agents and biofertilizers.

In forestry, including agroforestry, the efficacy and cost-effectiveness of techniques for micropropagation for various purposes should be critically analysed. In some species, such as poplars, biotechnology for pest and disease management is an attractive proposition.The commodities of high food and non-food values in local settings, but of little economic significance to the capital-intensive markets of industrialized countries, often referred to as "orphan" commodities, should receive due attention from local biotechnologists. Commodities such as coconuts, oil palm, pigeonpea, jute and buffaloes are almost monopoly commodities of the region and responsibility for their biotechnological improvement should fall primarily to the countries of this region.

Each country should establish national biotechnology committee comprising government agencies, universities and scientific academies, the mass media, industry and financial institutions - into a symbiotic relationship. The public sector in several Asian countries has allocated considerable budget and other resources and has further plans to strengthen biotechnology research and development. Since these are relatively costly fields of research, and recognizing that the indigenous capacity in frontline research is essential for sustained progress in application of the new technologies, governments must allocate adequate budgets to biotechnology and ensure the most judicious utilization of funds according to well-chosen priorities. The private sector is also active in biotechnology in some countries. Appropriate policies and measures must be established to promote private-sector involvement in this field and to forge synergistic links between the private and public sectors.

A few regional cooperative networks on biotechnology are operational in the region, and more may be initiated in the future. Usually these networks are supported through external funding and often collapse with the termination of the external support. For instance, after the regional animal biotechnology project was terminated in 1993, very little regional collaboration exists in this field. Mechanisms to sustain such activities should be established. One such mechanism may be the continuation of some selected activities under the umbrella of the Asia-Pacific Association of Agricultural Research Institutions (APAARI). Member countries should specifically contribute to agreed activities and a regional working group on biotechnology could be established under APAARI.

Notwithstanding its high potential, biotechnology should be seen essentially as a tool to complement the efficacy and effectiveness of conventional approaches to solve problems already on the agenda. Countries should have mechanisms to decide priorities and a blend of the most appropriate technologies to attain the goals. The temptation of "quick fix" and using/promoting a novel technology just for the sake of the technology should be avoided. On the other hand, appropriate capabilities, policies and infrastructures must be created in each country to exploit new and emerging technologies judiciously and rationally and not miss new opportunities.

Chapter 11

Agricultural Market Trends in India

Use of Chemical Fertilizers

After independence the use of fertilizers in India in the last 50 years has grown nearly 170 times. In 1950 use of fertilizer per hectare in India was 0.55 Kg but by 2001-02 this figure has increased to around 90.12 Kg per hectare. Green revolution during 1960s and subsequent increased intensification of agriculture were major causes behind this growth as seen in Fig. 1. Fertilizers and pesticides have become major cost of production in India along with the cost of other input like seeds, and labor cost.

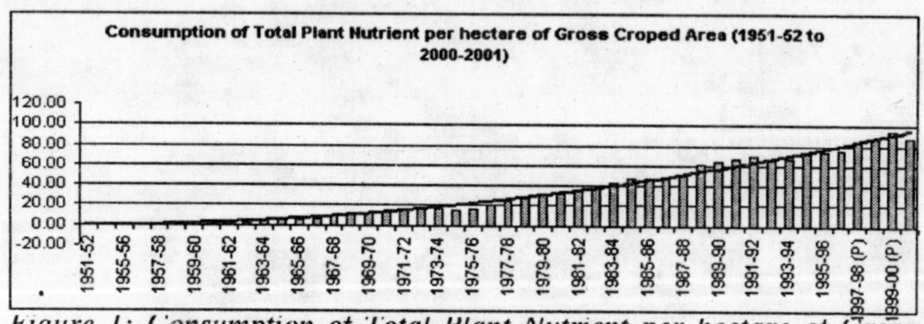

Figure 1: Consumption of Total Plant Nutrient per hectare of Gross Cropped Area

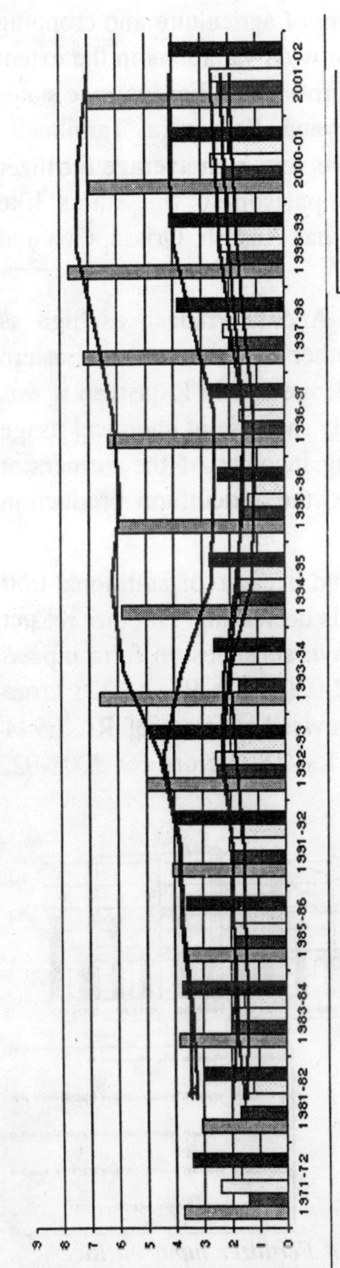

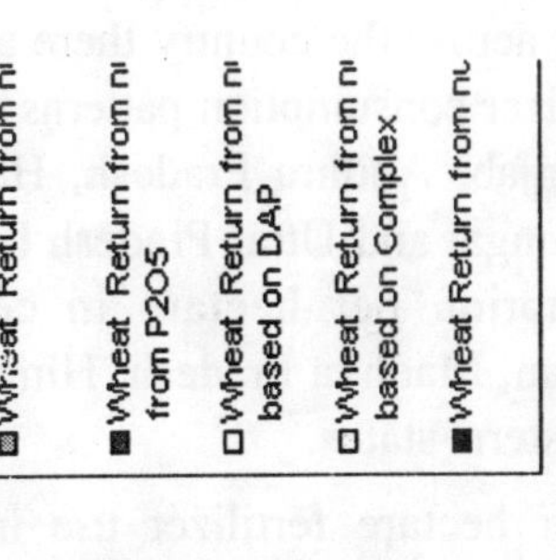

Figure 2: Trends in Economics of Fertilizer Input on Wheat Production in India (1971-2002)

Given the differences in the intensity of agriculture and cropping patterns across the country there are wide variations in the extent of fertilizer consumption patterns across India. For instance states like Punjab, Andhra Pradesh, Haryana, Karnataka, Tamilnadu, West Bengal and Uttar Pradesh have very high average fertilizer consumption per hectare in comparison to the states like Rajasthan, Madhya Pradesh, Himachal Pradesh, Orissa, Goa and northeastern states.

Per hectare fertilizer use in Andhra Pradesh as high as 179.2 Kg in 2000-01 and on the other hand in most of Eastern states it was less than 10Kg per hectare and in Rajasthan it was only 29.78 kg/hectare. Unfortunately increase in chemical usage has not inevitably led to increasing incomes of the farmers at the same rate at which it enhanced the agricultural production costs (Fig. 2 & Fig. 3).

The marginal income from land because of additional unit of chemical fertilizer and pesticide is decreasing. Another reason for increasing cost is decreasing govt. subsidies on farm inputs. The fertilizer subsidy bill for 2002-03(BE) is Rs. 11,228 crore that is Rs.716 crore less than the revised estimate of Rs. 1,944 crore and budgeted expenditure of Rs12,808 crore for 2001-02.

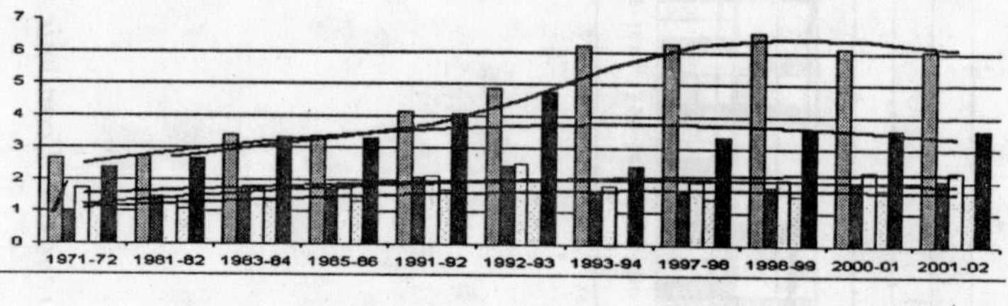

Figure: 3 Trends in Economics of Fertilizer Input on Rice Production in India (1971-2002)

Use of fertilizer in improper ratio of N: P: K is also a major problem for Indian agriculture. While the recommended ratio is 4:2:1 this ratio of N: P: K has been around 8.5:2.6:1. This "induces initial vegetative growth, susceptible to pests, diseases, lodging and causes poor floral induction and delayed maturity thereby reducing the yield".

Pesticides

Other major input for Indian agriculture is use of various pesticides, like insecticides, weedicides, fungicides, rodenticides etc. As the cropping pattern is becoming more intensive use of these pesticides is also increasing. Consumption of insecticide in agriculture has been increased more than 100% from 1971 to 1994-95. For instance, insecticide consumption in India, which was to the tune of 22013 tonnes has increased to 51755 tonnes by 1994-95. Consumption of all of these pesticides in same duration has increased more than two times, that is from 24305 tonnes to 61357 tonnes.

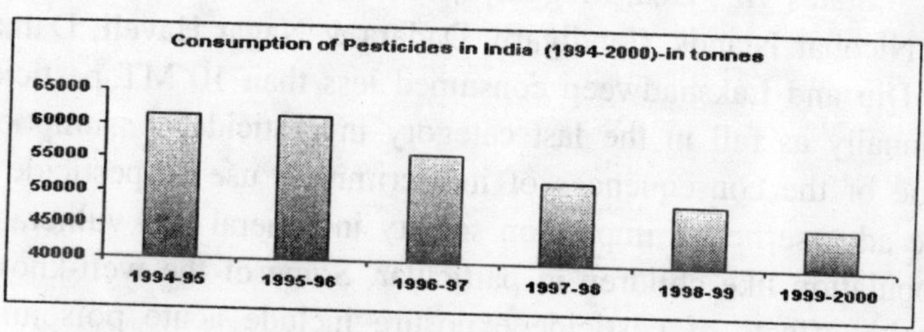

Figure 4: Consumption of Pesticides in India

But in recent past, change has been observed in trends of pesticides consumption. As a consequence of adoption of bio intensive Integrated Pest Management Programme in various crops the consumption of chemical pesticide (Tech. Grade) has come down from 66.36 thousand MT during 1994-95 to 43.59 thousand MT during 2001-02 with a reduction of 27.69%. Consumption pattern of pesticides in India is also very different from world. In India insecticide account for 76% of the total domestic market while herbicides & fungicides have a significantly higher share in the global market. There are wide ranges of regional variations in pesticide consumption in the country.

In the year 2000-01, States of Haryana, Punjab and Uttar Pradesh by consuming more than 5,000 MT (technical grade) pesticides annually come under category I state in consumption of pesticides. States viz., Andhra Pradesh, Gujarat, Karnataka, Kerala, Madhya Pradesh, Maharashtra, Rajasthan, Orissa and Tamil Nadu, which consumed between 1000 MT and 5000 MT fall in the category II states. States viz., Assam Bihar and Himachal Pradesh that consumed pesticide between 100 and 1000 MT come under category III. States viz., Arunachal Pradesh, Jammu & Kashmir, Manipur, Mizoram, Nagaland, Tripura, Delhi and Union Territory (UT) of Pondicharry consumed pesticides between 10 and 100 MT annually fall under category IV.

States viz., Goa, Meghalya, Sikkim and UTs viz., Andman & Nicobar Islands, Candigarh, Dadara & Nagar Haveli, Daman & Diu and Lakshadweep consumed less than 10 MT pesticide annually as fall in the last category in pesticide consumption. One of the consequences of indiscriminate use of pesticide is the adverse health impact on society in general and vulnerable population like children in particular. Some of the well-known health effects of pesticide exposure include acute poisoning, cancer, neurological effects and reproductive and developmental harm. The major causes of concern in this respect are bio-accumulation of pesticides and the prolonged time period that it takes to express the negative health consequences.

GREEN INPUT MARKET

It is very difficult to estimate the total size of green inputs market in India because of its diversity in terms of products and also due to the nature of it being unorganized market. Green inputs into agriculture include bio-fertilizers, bio-pesticides, compost, Farm Yard Manure (FYM), green manure etc. As most of this inputs are either not traded and even if they are traded, it is only at informal levels available information regarding production capacity, demand and sales is at best sketchy estimation and hence inadequate. The limited number well-established firms that have their presence in this market today as a result it this is predominately controlled by the small and local producers of bio-fertilizer, vermi compost and other input producers who are in large numbers.

Relatively speaking there is more rigour in estimation of bio-fertilizer market in India, because of the presence of some large producers in production of bio-fertilizers and comparatively this is more organized than other green inputs market. Based on the gross cropped area in India (190 million hectare) and recommended doses of bio-fertilizers, potential demand is estimated to be about 6,27,000 MT. This demand can further be segregated into different categories of bio-fertilizer, such as Rhizobium, Azotobactors, Azospirillium, BGA, and Phosphate Solubilizer etc. whose demand differ widely as shown in Table 1 But by looking at the product segregation as seen in Table:1 potentials for different strains of bio - fertilizers in India:

Table 1: Estimated Total Potential Demand for Bio-fertilizers in India

Category of Bio- Fertilizer	Amount in Million Tonnes
Rhizobium	35,730Mt
Azotobacter	162,610Mt
Azospirillum	77,160Mt.
BGA	267,510Mt
Phosphate solubilizer	275,510Mt
TOTAL	818,730Mt

For year 2000 proposed production target for bio-fertilizer was 39,165Mt, which was only 4.8% of the total estimated demand. But the actual production and the distribution of bio fertilizers are below the targeted (Table.2). This shows the huge gape between potential market demand and production and also provides an opportunity for bio-fertilizer producers.

There has been a positive trend in India, with respect to production of bio-fertilizers. As it can be seen from Table 2 that while the total production of bio-fertilizer in India in 1992-93 was 2005.0 tons and it has increased to 8010.1 tons by 1998-99. Similarly the consumption/distribution of bio-fertilizer has also increased 1600.01 tons to 6700.27 tons during the same time period.

Table: 2 Installed production capacity, total production and distribution of bio-fertilizer in India (1992-99)

Year	Installed Production Capacity (Tons)	% Growth Rate in Installed Capacity	Total Production (Tons) %	Growth Rate in Production	Total Consumption/ Distribution (Tons)	% Growth Rate in Consumption/ Distribution
1992-93	5400.5		2005.0		1600.01	
1993-94	6125.5	13.42	3084.0	53.82	2914.37	82.15
1994-95	8114.5	32.47	5800.5	88.08	4988.90	71.18
1995-96	10680.4	31.62	6692.3	15.37	6288.32	26.05
1996-97	12647.0	18.41	7406.6	10.67	6681.44	6.25
1997-98	N.A.	0.00	7104.6	-4.08	6295.63	-5.77
1998-99	16446.0	30.04	8010.1	12.75	6700.27	6.43

But the growth rate in installed bio-fertilizer capacity is comparatively more stable than the growth rate in total production, consumption & distribution of bio-fertilizers (Table: 2). For total production of bio-fertilizers growth rate had reduced to 12.75% during 1998-99 from 53.82% in 1993-94 and during same period the growth rate of consumption & distribution had reduced to 6.43% from 82.15%. This is an indication that there is not only a need but also a role for market development for green agriculture inputs in India.

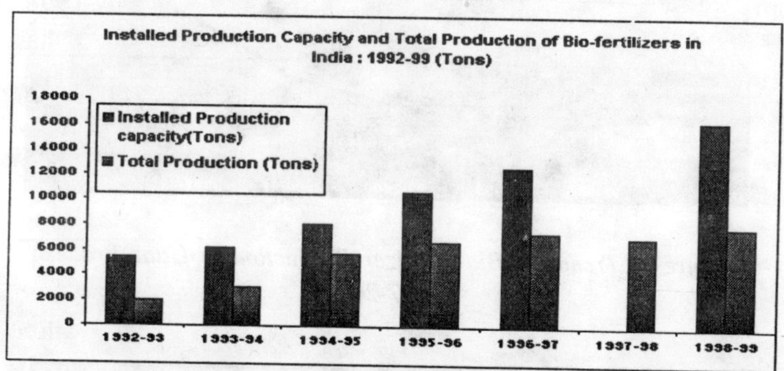

Figure: 5 Installed production capacity and total production of bio-fertilizer in India (1992-99)

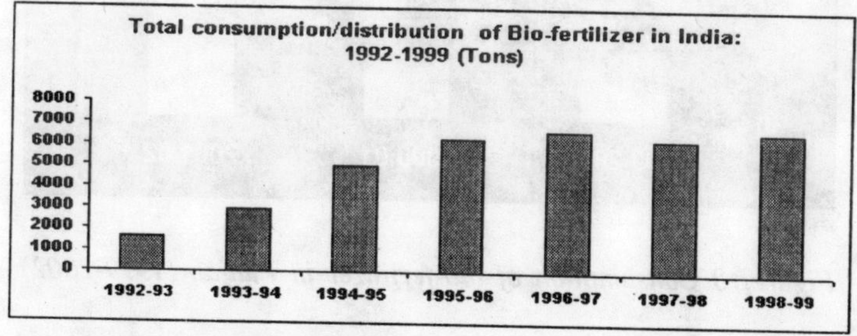

Figure: 6 Total Consumption / Distribution of Bio-fertilizer in India (1992-99)

Inspite of the impressive growth rate of more than 200% in production capacity and around 300% growth rate experienced in production and consumption of bio-fertilizers in India within 6 years (i.e., from 1992-93 to 1998-99) yet it is only around 1.5% of the estimated demand potential for bio-fertilizers in the country.

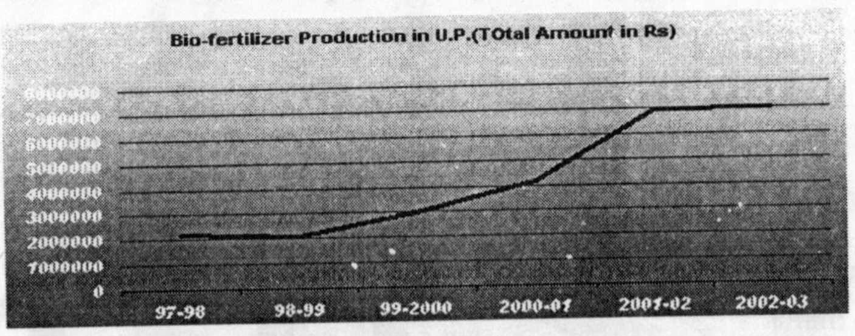

Figure 7: Trends in Bio-fertilizer Production in Uttar Pradesh (1997-2003)

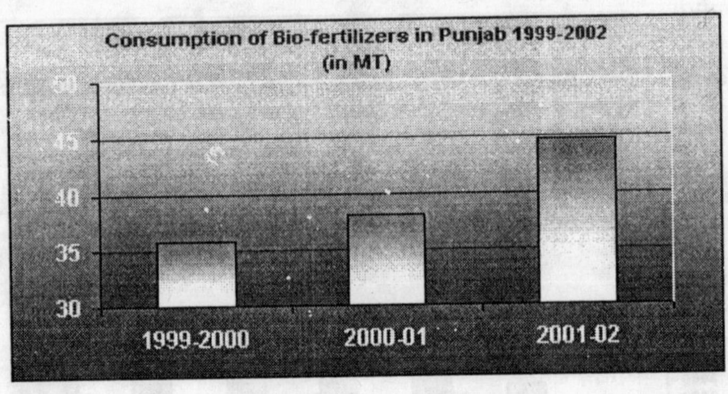

Figure: 8 Consumption of bio-fertilizer in Punjab (1999-2002)

Region wise production trend analysis in India shows wide variations across the country. Western India has the highest bio-fertilizer production capacity as it can be seen that for the 1999-2000 the estimated production in this region was 5098.2 tons and it was expected grow upto 6429.9 tons by 2000-01. Southern

India has the second largest bio-fertilizer producing region with 4491.6 tons for 1999-2000 and 6117.5 tons for 2000-01. Inspite of the huge agricultural production base in Northern India the production of bio-fertilizer stood at 207.2 tons for 1999-2000.

At state level Maharashtra is highest bio-fertilizer producer state with 3173.0 tons of likely production for 2000-01 followed by Karnataka with likely production of 3152.5 tons for same year (FAI, 2001). Other states are also showing the trends of increasing orientation towards bio-fertilizers. For example in last four years value of bio-fertilizer distributed in Uttar Pradesh has increased from around Rs 2 Lakh in 1997-98 to Rs. 7 Lakh in 2001-02 (Fig: 7). Punjab is also showing increasing trends for bio-fertilizer market inspite of the fact that the sale of bio-fertilizer in Punjab was started in only 1997-98 by 2001-02 market has grown to 45 mt/year (Fig: 8).

Increase in use of vermi-compost has been observed for kitchen gardens and for cultivation of high value cash crops but the information about actual quantity used is not available. Area under green manure is one indicator that is showing negative trends because the area under intensive cropping is increasing. Increase in irrigation facilities are indirectly contributing towards reduction of area under green manure is as seen in Table 3.

Table: 3 Area under Green Manure in India (1995-97)

Year	Area in Lakh Hectares
1994-95	35.872
1995-96	34.411
1996-97	22.512

Other green inputs for agriculture in India are used in very minimal quantity. Some of the popular bio-pesticides include neem based formulation and Bacillus thuringingiensis (Bt). Consumption of bio-pesticides in India has increased from 83 MT (Tech. Grade) during 1994-95 to 686 MT during 1999-2000 and in case of neem based pesticide formulation it has increased 40 MT to 71 MT during the same period in case of Bacillus thuringiensis (Bt).

GREEN OUTPUTS MARKET TRENDS

Demand for green agricultural products is a stimulant for growth for input market. In other words if there is demand in market for organically produced farm products will encourage farmers to implement the organic farming practices and also to use organic input like bio-fertilizers, bio-pesticides, varmi-compost, green manure and FYM.

Estimating area under organic agriculture in India is a difficult task as there is no central agency that collects and compiles this information. Different agencies have estimated the area under organic agriculture differently for instance the study undertaken by FIBL and ORG-MARG the area under organic agriculture to be 2,775 hectares (0.0015% of gross cultivated area in India). But other estimation undertaken by SOEL-Survey shows that the land area under organic cropping is 41000 hectare. The total numbers of organic farms in the country as per SOEL-Survey are 5661 but FIBL and ORG-MARG survey puts it as 1426. Some of the major organically produced agricultural crops in India include crops like plantation, spices, pulses, fruits, vegetables and oil seeds etc (Table: 4).

Table: 4 Major products produced in India by organic farming

Type of Product	Products
Commodity	Tea, Coffee, Rice, Wheat
Spices	Cardamom, Black pepper, white pepper, Ginger, Turmeric, Vanilla, Tamarind, Clove, Cinnamon, Nutmeg, Mace, Chili
Pulses	Red gram, Black gram
Fruits	Mango, Banana, Pineapple, Passion fruit, Sugarcane, Orange, Cashew nut, Walnut

Vegetables	Okra, Brinjal, Garlic, Onion, Tomato, Potato
Oil seeds	Mustard, Sesame, Castor, Sunflower
Others	Cotton, Herbal extracts

India is best known as an exporter of organic tea and also has great export potential for many other products. Other organic products for which India has a niche market are spices and fruits. Org-Marg's survey also proves this fact as around 30% of respondents that includes producer, exporters and traders has responded that organic tea is produced in India and this is highest response for any single crop, next are spices, fruits, vegetables, rice and coffee. There is small response for cashew, oil seed, wheat and pulses. Among the fruit crops Bananas, Mangos and oranges are the most preferred organic products.

Export Market: Organic agricultural export market is one of the major drivers of greening of agriculture in India. The current production of organic crops is around 14,000 tons. Out of this production, tea and rice contributes around 24% each, fruits and vegetables combine makes 17% of this total production. From India around 11,925 tons of organic product is exported, that makes around 85% of total organic crop production. Major export market for Indian producers are Australia, Belgium, Canada, France, Germany, Italy, Japan, Netherlands, Sweden, Singapore, South Africa, Saudi Arabia, UAE, UK, and USA. Estimated quantity of various products that are exported from India in 2002 is shown in Table 5.This shows that around 3000 tons of tea was exported and in quantity term it was the highest, next major exports are rice (2500 tons), fruits & vegetables (1800 tons), cotton (1200 tons) and wheat (1150 tons).

Table: 5 Major organic products exported from India

Product	Sales (Tons)
Tea	3000

Coffee	550
Spices	700
Rice 2500	
Wheat	1150
Pulses	300
Oil Seeds	100
Fruits & Vegetables	1800
Cashew Nut	375
Cotton	1200
Herbal Products	250
Total	11,925

The burgeoning US and European green markets provides enormous scope for Indian exporters. International Trade Centre's (ITC) overview of estimation of organic food world-wide shows that there is strong growth in retail sales from US $ 10 billion in 1997 to US$ 17.5 billion in 2000 and about US$ 21 billion in 2001 (in 16 European countries, USA and Japan). If the demand of so called 'green product' in Japan that is not certified as organic product is excluded from total estimation, than also it is US$ 16 billion for 2999 and reached around US$ 19 billion in 2001.

Though the current market share for organic produced is estimated between one to two percent of total food products market, but the forecasting by experts show that this market is likely to grow at a higher pace. According to experts in medium term around five percent of the market is expected to be organic market share. Europe is largest market of organic produces in world and consumes around half of the world produce of organic production. EU is a net importer of cereals, oilseeds, potatoes and vegetables. For 2001, European market for organic food was estimated to be around US$ 9 billion but with expected annual growth rate of around 20% (Table 6) depending upon the market, and for 2003 the retail sales for organic food in this market is expected to US$ 10-11billion.

Within Europe, Germany is largest market for organic products with sales value of around 2.5 billion Euros ($2.3 billion (US)). On an average the per capita spending on organic produces in Europe was euro 23 for 2000. In terms of per capita consumption of organic products countries like Denmark (72 Euros per head), Switzerland (Euros 68), Austria (Euros 40) and Germany (Euros 31) fared much better than others (Fig: 9). While some of the European countries like Italy not only meet their internal demand for organic products but they also cater to the demands of other neighbouring countries. There are other countries like United Kingdom, which are highly dependent on imports to meet their domestic organic product demand.

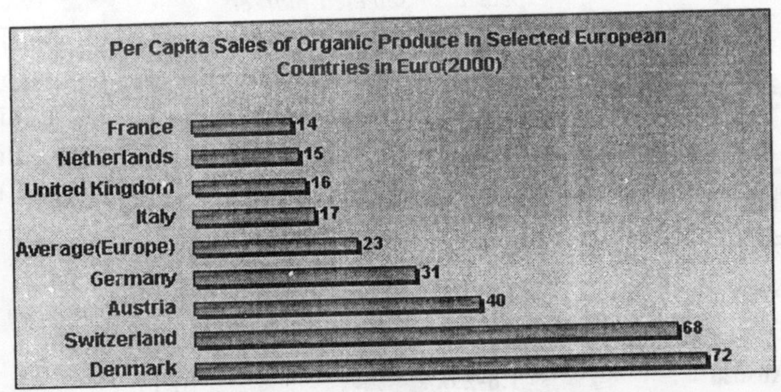

Figure 9: Per capita sales of organic produces in selected European Countries in Euro (2000)

In North America, retail sales of organic products for 2002 was estimated nearly $12 billion (US) out of this US alone is contributing for $11 billion (US). US retail sale for organic product has grown 20-24% per year for the past 12 years and the same growth trends expected to continue for future. Current retail sales for organic food is around 2% of total retail food sale in US.

In Asia largest market for organic food product is Japan and it is estimated to have the retail sales of organic food and beverage of around $(US) 2.5-3.0 billion (Minou Yussefi and Heldge Willer, 2003). Of this total value of organic market, imports are estimated to be to the tune of around $(US) 360 million. Though the organic food market is not more than 0.5% of total food market of Japan but according to the Japanese Integrated Market Institute, import of organic products is likely to grow by 40%. Other global markets for organic products in Middle East are Saudi Arabia and UAE. Within Africa, South Africa is the only country that has organic market potential. As seen from the global market growth trends for organic foods there are enormous potentials for India to exploit.

Table: 6 Percentage of organic food and medium term growth expected in selected markets

Markets	% of total food sales	% Expected growth - Medium term
Germany	1.6-1.8	10-15
U.K.	1.0-2.5	15-20
Italy	0.9-1.1	10-20
France	0.8-1.0	10-15
Switzerland	2.0-2.5	10-15
Denmark	2.5-3.0	10-15
Austria	1.8-2.0	10-15
Netherlands	0.9-1.2	10-20
Sweden	1.0-1.2	15-20
Belgium	0.9-1.1	10-15
U.S.A.	1.5-2.0	20

Total commodity wise demand has been estimated in some selected export markets (Germany, Holland, UK, Switzerland, USA, and Japan) by the FIBL & ORG- MARG survey which shows that for Banana it is 6,410 tons, for wheat and soy bean 1,000 tons, for pineapple around 900 tons and for mango this is around 650 tons.

The attractiveness of organic market gets enhanced also because of the price premium that these products have over the conventionally produced products. Price premium for various organic produces vary in different countries depending upon the distribution channels and market conditions. This premium varies from 30-50% (trader level) for different products.

As seen from above there are immense opportunities for organic agricultural exports for Indian to exploit. Some of the prerequisites for exploiting this potential include:

— Farmers capacity to produce the organically the agricultural products which have global market and

— Prior experience of exporters and traders in exporting agricultural commodities to these markets

In the Fig 10 an attempt has been made to develop a matrix by depicting the conventional agricultural commodities, which India has been exporting to various countries across the world as well as the existence of organic market for these commodities in these countries. Depicting the current conventional agricultural market indicates the capabilities of India in exporting the specific agricultural commodities to different countries and similarly depicting the existence of organic agriculture market for specific commodities in these countries reveai the existence of opportunities for exporting organic agricultural commodities. In developing this matrix we have used annual exports of agricultural commodities as published in CMIE agricultural sector reports and for exploring organic market in different countries for different commodities we have used the internet resources mentioned at the end of this article.

The matrix reveals that India has demonstrated capabilities of exporting agricultural commodities like rice, wheat, tea, coffee, spices, oil meals, sugar, fruits & vegetables etc to countries like USA, U. K, Germany, Japan, France, Saudi Arabia, South Africa, CIS Countries, Poland, Netherlands, Italy etc. It also shows that in most of these countries there is a demand for organically produced commodities, which attract price premiums ranging from 10% to 100%. This showcases a window of opportunity that is yet to be exploited to its maximum potentials by Indian exporters and producers of agricultural commodities.

Agricultural Commodities	Rice	Wheat	Tea	Coffee	Tobacco	Spices	Cashew	Oil meals	Cotton	Castor oil	Sugar	Fruits / vegetables
USA	❀		❀	❀	❀						❀	
UK	❀		❀		❀	❀	❀					❀
Japan			❀	❀		❀	❀	❀				
France						❀	❀				❀	
Germany			❀	❀	❀	❀	❀					❀
Italy				❀								
CIS	❀		❀	❀		❀	❀					
Netherlands			❀	❀			❀					
Switzerland				❀								
Egypt			❀		❀						❀	
UAE	❀		❀		❀	❀	❀					❀
Saudi Arab			❀		❀							❀
Poland			❀	❀								
Belgium				❀	❀							
South Africa	❀											

❀-Existing conventional export market for Indian producers for particular product

[shaded box] Prospective market for Indian organic products.

Figure 10: Conventional agricultural products & their export market and prospective market for Indian organic products

Domestic Market: Given the unorganized nature of the domestic organic agriculture market in India it is difficult to estimate the magnitude and trends in this growing market. In general, the sale of organic produces is limited to metros like Mumbai, Delhi, Kolkata, Chennai, Bangalore and Hyderabad. To a large extent this sale is also based on individual initiative of the farmers, Non Governmental Organizations and some entrepreneurial traders etc. The current domestic green products market demand is mainly for fruits, vegetables, rice and wheat. Other products include tea, coffee and pulses.

The market prospects other commodities like organic spices, fruits, herbal plants and cotton are relatively high. For next five years it is projected that organic spices would grow by 14%, fruits 8% and herbal plants and cotton it is estimated to be around 7%. Market for different range of organic agricultural products as shown in Table 7 is estimated to reach up to 1568 tons in 2006-07.

Table: 7 Growth forecast for specific organic products in the domestic market

Product	% Projected Growth in the 5 next Years
Spices (all)	14
Pepper	5
Turmeric	4.5
Tea	13
Rice	10
Fruits (all)	8
Banana	15
Mango	5
Orange	5
Pineapple	5
Herbal extracts	7
Cotton	7
Coffee	5
Oil seeds	5
Honey	5
Groundnut	5
Baby food	5
Coconut	5

GOVERNMENT'S ROLE IN PROMOTION OF GREEN AGRICULTURE

There has been limited allocation under 9th Plan for bio-fertilizer, bio-pesticide popularization in the Agriculture sector and the provision for this programme was enhanced by expanding the scope of activities whereby promotion organic agriculture has been included as part of the activities targeted in the 10th Plan. As part of 10th Five year Plan, Government of India has earmarked about Rs. 100 crores for the promotion of organic agriculture in the country. The main components of this initiative include farming of standards, negotiating with different countries and putting in place a system of certification for organic products.

Central Government is also promoting the production and use of bio-fertilizer to make it popular. Government has initiated a project "National Project on Development and Use of Bio fertilizers" for this purpose. Main objectives of this project are as following:

i. Production and distribution of Bio fertilizers (BFs)

ii. Developing Standards for different BFs and Quality control

iii. Releasing of grants for setting up BF units

iv. Training and Publicity

Promoting green agriculture market: To promote the organic agriculture in India government has also taken some initiative in recent past. APEDA (Planning Commission, 2001) is the nodal agency to promote the Indian organic agriculture and its exports opportunities. National Steering Committee under the Chairmanship of Secretary Commerce has already outlined and approved the National Program for Organic Production (NPOP) by May 2001. Under this program, National Organic Standards have been evolved. This apart, it has also developed Criteria for Accreditation of certification agencies, Accreditation Procedure and Inspection and Certification Procedures.

In developing these standards and procedures due attention is paid to the guidelines as enumerated by international organizations such as International Federation for Organic Agricultural Movement (IFOAM), EU Regulations and FAO Codex Standards. As part of this program, a National Logo for organic products on behalf of Govt. of India has also been developed. Some of the other efforts towards promotion of organic exports include attempts to collaborate with all the major organic importing countries. Towards this APEDA is deliberating with European Union for inclusion of India in the list of third countries under Article 11 of the EU regulations No 2092/91 so that India's National Programme for Organic Production gets the required recognition under the EU regulations.

Facilitating Factors

Organic agriculture provides economic opportunities for different stakeholders. Some of the drivers that facilitate growth of organic agriculture in India are

— Growing export market for organically produced crops.

— Price premium for organically produced agriculture products from10% to 100%.

— Diverse agro-climate regions across the country that provides environment for wide range of crops that can cater to different market demands.

— Increasing awareness & health consciousness especially among certain sectors of domestic consumers.

— Availability of comparatively cheap labor for labor-intensive organic agriculture

— Huge numbers of small farmers those who do the traditional farming with very limited capacity to pay for most of the chemical inputs into agriculture

— Presence of Non-Government Organizations (NGOs) as active promoters of Organic farming in different agro-climatic regions

— Increasing involvement of private companies in field of agricultural extension, trade, consultation and other services

— Enhanced Government attention and support for organic agriculture through various policy initiations and action programs.

Though there are positive signs for green agriculture in India it is not growing at a pace to enhance its market attractiveness so as to motivate larger section of farming community to opt for organic agriculture. Fig 11 is an exercise to find why the penetration of greening the agriculture in India is limited. Major problems that hinders the growth organic agriculture in India can be listed as follows:

— Lack of proper infrastructure for distribution and conservation of bio-inputs is a major constraint hinders the access of these inputs to farmers.

— Existence of poor quality bio-inputs in market reduces the credibility of input providers. Lack of quality control mechanisms for bio-inputs furthers the mistrust among farmers.

— Given the low penetration of bio-inputs market and the limited shelf-life it is disincentivizing the traders to store and sell bio-inputs

— Bio-fertilizers and bio-pesticides are perceived as less yielding.

— Some climatic regions and soil conditions are not suitable for specific strains of organic production.

— For some strains limited shelf life is also constrain as most of the bio-input last only for about 4-6 months.

— Given the mandated gestation period of around three years for a conventional farm to become an organic farm the benefits perceived by farmers in general and small and marginal farmers in particular tend to be limited as they have short term orientation. As a result even if they are aware they are hesitant to switch over to organic (green) agricultural practices.

— Agricultural departments, research institutions and extension services have for long been oriented towards chemical input agriculture as a result there is a requirement for reorienting these officials towards organic (green) agriculture

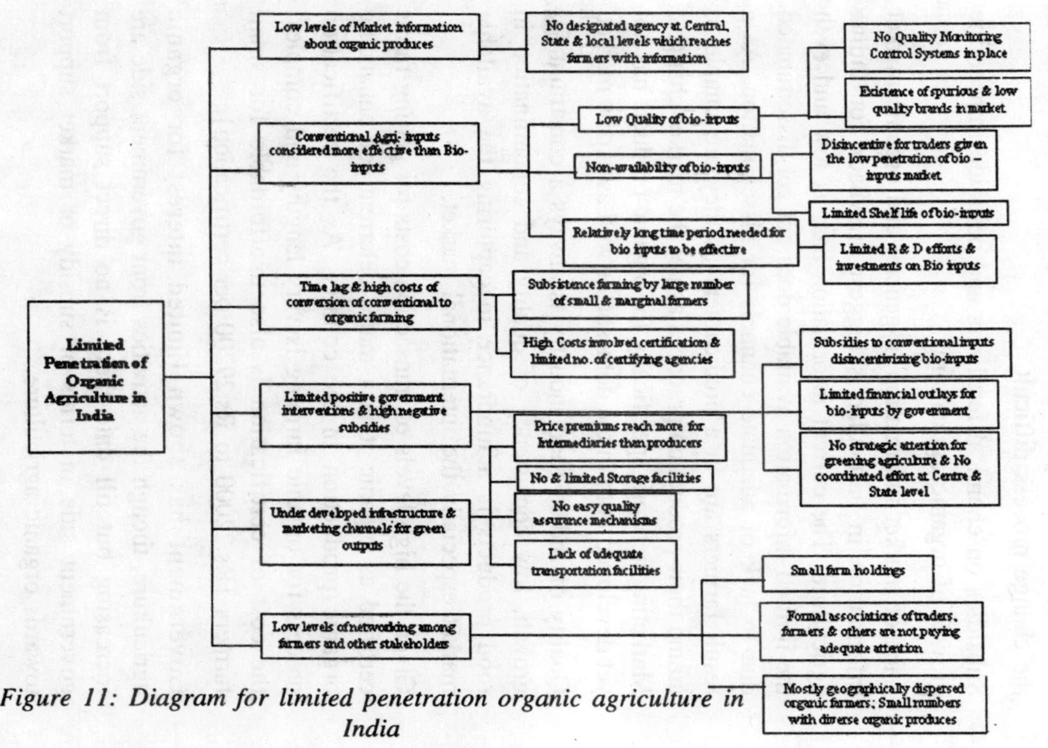

Figure 11: Diagram for limited penetration organic agriculture in India

— Changing the cropping and cultivation patterns is slow and time-consuming process given the high levels of illiteracy and large number of small and marginal farmers it makes the change process difficult.

— Subsidies on chemical fertilizers and pesticide impede the growth of organic agriculture.

— Lack of market information in general and organic market information in particular is biggest drawback for Indian agriculture. The current information base is low and even the limited information available does not get disseminated due to lack of adequate channels for dissemination. As a result farmers are in a predicament as they are unable to attune their production practices as per the market changes. Marketing network specifically for organic products has not yet developed both in the domestic as well as export market.

— Quality of Indian food industry is always a constraint for growth, low consistency of quality and contamination in food products is a hindrance in capturing the available market especially the international market.

— Given the high levels of transaction costs for getting farms certified as organic it is a major deterrent for enhancing organic production in the country. As the certification process for organic farming is very lengthy and complex, the cost of certification is also unaffordable for small farmers {Rs.22000 to Rs.29200 per certification}.

— Government has shown limited interest for organic agriculture, though the activities from government side are increasing but till date there is no direct support from government side in terms of subsidy or market support towards organic agriculture.

— Lack of proper infrastructure in terms of roads from remote villages, cold storage facilities and slow transportation infrastructure affects the cost, quality and reach of producers and

— Indian organic agriculture is very fragmented and there are no organizations for managing the entire value chain of organic products.

Large number of these problems listed above are due to the relatively newness of this sector from the point of view of different players.

Experience elsewhere shows that government has to play a key role in developing organic agricultural production and enhancing marketing opportunities. Towards this there is a need to have policy framework to address greening agriculture in India. Policy change in favour of organic agriculture can make positive difference for changing the market condition in terms of encouraging production of bio-inputs, which in turn can propel changes in cropping pattern in favour of organic practices. Currently the attention given both in terms of policy framework and institutional dynamics towards organic agriculture is only marginal. Though Ministry of Commerce registers farmers wishing to convert to organic operations but farmers are asked to seek technical assistance from the Ministry of Agriculture that is unavailable.

Involvement of government not just in standardization and accreditation procedures but also through proactive "support to inspection and certification and market-oriented services are necessary to provide equal opportunities. Otherwise, the export of certified organic products risks becoming that only large farmers, or highly organized groups of small holders, can afford".

Recommendations from Farmers Perspective:

— Existing subsidy structure enhances the use of chemical fertilizers and pesticides as agricultural inputs. Gradual phasing out subsidy on synthetic fertilizers / pesticides and initiating promotion of bio-inputs would contribute towards enhancing organic agriculture production in the country.

— Providing grant assistance in the form of financial assistance for converting the traditional farm in to organic farm and intending support to meet certification fees especially for the small and marginal farmers etc.

— Enlarging the scope of work of the agricultural extension services by focusing on organic agriculture through collaborative engagement with NGOs, who are actively working in the rural areas.

— Special insurance scheme for organic farm that reduces the risk of farmer in case of failure of crop.

— Promote contract farming based on organic agriculture as it has been done in Madhya Pradesh.

From NGO's Perspective:

— Non-Government Organizations have being playing crucial role in promoting organic agricultural practices in the country. Change of agricultural practices is equivalent to changing the culture and mindset of farmers, which can only be achieved by a long drawn interface. NGOs have demonstrated capabilities to this effect. An illustrative example of public private partnership is the successful story of Spice Board's involvement of NGOs to enhance organic production of spices in Kerala, Tamilnadu, Andhra Pradesh and North Eastern states.

From Traders Perspective:

— Ministry of Commerce & Trade should have a dedicated cell for providing organic agricultural export market potentials in different parts of the world and the price premium that different products attract.

— Establishment of Special Organic Agriculture Trade Zone (OATZ) for domestic as well export market can help the traders in easily accessing the farm products as well as customers for organic products like agro- based food processors and retailers.

— In order to develop organic export networking potentials there is need for inventorizing the early exporters by providing special tax relaxation for traders/exporters those who have organic trading/export.

— In order to enhance the domestic organic agricultural market government need to make efforts towards promotion of organic food processing industries by providing subsidies and financial assistance as well as facilitates to market development. As the shelf life of some of the organic produce is shorter establishment of dedicated cold storage facilities could help in enhancing the prospects of exports.

From Green Input Producers Perspective:

i) Popularization of existing scheme for promotion of bio-fertilizers and extension of these schemes for other bio-inputs.

ii) Assistance in marketing of bio-inputs thorough govt. network and also involving the network of co-operative societies at village level for distribution of these bio-inputs.

iii) Improving infrastructure facilities like roads, transportations, storage facilities etc would enhance forward and backward linkages in the organic products supply chain.

iv) Promotion of corporate research for organic agronomic practices, bio- control of diseases and pest, bio-fertilizers etc.

From Institutional Perspective:

There is urgent need for giving strategic attention to organic farming efforts. This requires inter ministerial coordination among various ministries at the Centre. Towards this forming a Steering Committee consisting of various Ministries at Central

Government level would be a move in the right direction. Such as..

— Ministry of Agriculture,

— Ministry of Commerce & Trade,

— Ministry of Environment,

— Ministry of Science & Technology,

— Ministry of Finance etc.

Such committee would helps on increasing effectiveness of policies & programmes evolved towards promotion of organic agriculture.

At the state level some of the important institutions that require coordinated action plan include: Agriculture universities, State agriculture department, Private business organizations and NGOs. Each of these institutions can enhance their contribution towards greening agriculture by developing a detailed program of action.

Chapter 12

Perspectives in International Agricultural Research

The Consultative Group on International Agricultural Research (CGIAR) is a global agricultural research network, and the largest scientific partnership in its field. Its mission is to "contribute to food security and poverty eradication in developing countries through research, partnership, capacity building, and policy support, promoting sustainable agricultural development based on the environmentally sound management of natural resources. "

The CGIAR has attempted to use new developments in molecular biology—including tissue culture, gene markers in traditional breeding, and genetic engineering—as elements in a comprehensive sustainable poverty alleviation strategy focused on broad-based agricultural growth. These efforts seek to integrate biotechnology with conventional and traditional approaches, and are carried out with due regard to the potential risks as well as the potential benefits.

Established in 1971, the CGIAR currently pursues these goals through the activities of 16 international agricultural research centers (IARCs). The Group's co-the U. N. Development Program (UNDP), the U. N. Environment Program (UNEP), and the World Bank. Among the 58 members of the Consultative group are governments of developing, transition, and industrialized countries; international organizations; and charitable foundations.

The IARCs affiliated with the system are probably best known for their role in the Green Revolution, i. e. , breeding high-yielding varieties of maize, wheat, and rice. These efforts played a pivotal role in tremendous increases in cereal yields that have occurred since 1960, as well as the 20 percent gain in daily per capita calorie availability. The CGIAR centers are also involved in research on other food crops consumed by poor people, including millet, barley, sorghum, root and tuber crops, bananas and plantains, and food legumes, as well as on livestock, aquatic resources, water resource management, and agroforestry. In addition, individual centers are devoted exclusively to policy research and strengthening the capacity of national agricultural research systems in developing countries.

The biological research centers also integrate such activities into their research programs. The 16 centers collaborate with one another on many projects and programs, covering such critical areas as property rights and collective action in natural resource management, integrated pest management, integrated natural resource management, farmer participatory research and gender analysis, alternatives to slash and burn agriculture, breeding micronutrient-rich food crops, and sustainable agriculture in mountainous areas.

From its inception, the CGIAR has focused not only on "increasing the pile of food" available to people in developing countries through technological improvements, but also on "ecological, economic, and social factors, " as stated in the Group's founding resolution. The emphasis on poverty reduction and sustainability has increased over time. Research at CGIAR centers addresses specific agricultural problems facing poor people in developing countries, such as how to intensify food production in less-favored areas in a sustainable manner, or how to assure that increasing demand for meat and milk products in developing countries can benefit low-income farmers and consumers.

It strives to join cutting edge-science with insights from traditional and indigenous knowledge, as well as agroecology. As a publicly funded network, financed by taxpayers from both the South and the North and by philanthropic institutions, the research outputs of the system constitute "international public goods. " Economists use the term public goods to characterize goods and services that one person can consume without detracting from the consumption of another. It is also impossible or very difficult to exclude anybody from consuming such goods and services.

New knowledge created by CGIAR centers is made freely available to all, including public policy makers, for the benefit of the global public, especially poor people, and can be used over and over again. Results from CGIAR research are made available in the public domain in keeping with their publicly funded, public interest, nonprofit character. The new knowledge and technology developed at IARCs is international in character because it is relevant to many countries. Presently, the CGIAR seeks to increase agricultural productivity in developing countries through genetic improvements in plants, livestock, fish, and trees, and through better management practices.

Increasing food crop resistance to insects and diseases is a major focus of CGIAR research, which utilizes a wide variety of techniques, including biological control, conventional intra-species crossbreeding, and various forms of biotechnology, such as tissue culture, molecular genetics, and genetic engineering. CGIAR research has also contributed to conservation of natural resources, especially soil and water, and reducing the impact of agriculture on the surrounding environment. By improving productivity on existing farmland, CGIAR-developed technologies have conserved an estimated 300 million hectares of fragile land over the past 30 years, an area equivalent to the United States, Canada, and Brazil combined. IARCs affiliated with the system have played a leading role in advancing integrated pest management strategies, utilizing biological control approaches and avoiding synthetic pesticide use as much as possible.

Such approaches are not only environmentally friendly, but accessible to small farmers who may not be able to afford chemical pesticides. Two CGIAR centers, the International Center for Tropical Agriculture (CIAT) in Colombia and the International Institute of Tropical Agriculture (IITA) in Nigeria collaborated on an effective biological approach (the introduction of a South American parasitic wasp) to cope with the cassava mealy bug, a pest that wreaks havoc with a crop essential to the diets of many poor Africans. It is estimated that the application of this research in Sub-Saharan Africa has led to increases in cassava production worth $400 million a year.

In addition, the system is a major player in efforts to save biodiversity. CGIAR centers' genebanks hold 600,000 accessions of more than 3,000 crop, forage, pasture, and agroforestry species. The collection consists of wild species, farmers' varieties, and varieties improved by crop scientists. It is one of the world's largest ex situ collections of plant genetic resources. In accordance with the Convention on Biological Diversity, the CGIAR has placed the collection under FAO auspices in trust for the global public. Duplicates of materials held in-trust are made available to researchers throughout the world without restriction or charge, so that new gene combinations can be used to address food problems. The common material transfer agreement used by the CGIAR centers in making in-trust germplasm available states that recipients may not apply intellectual property rights protection to the accessions received.

CGIAR centers are very active in efforts to preserve biodiversity in situ as well. For example, the International Plant Genetic Resources Institute (IPGRI) in Italy is working with farmers in Burkina Faso, Ethiopia, Hungary, Mexico, Morocco, Nepal, Peru, Turkey, and Vietnam to promote local exchanges of seeds and knowledge and to characterize, monitor, and conserve local crop biodiversity. This participatory, community-based approach allows farmers to collaborate directly with ex situ conservation efforts to prevent extinction of rare species. Following the genocide and mass exodus from Rwanda in 1994, CIAT and partner institutions from the agricultural research, governmental, and nongovernmental communities carried out the "Seeds of Hope" project. This multiplied Rwandan germplasm that had been stored at genebanks elsewhere, including CIAT, and restored the seed to Rwandan farmers. Since then, CGIAR centers have carried out similar projects in other countries affected by war and natural disasters.

Policy research is another major thrust of the CGIAR's work. In Bangladesh, the International Food Policy Research Institute (IFPRI) collaborated with the Ministry of Food on studies of the effectiveness of official grain procurement and the rural ration shop program. This work determined that private grain marketing would save the government $37 million without harming the food security of poor Bangladeshis. In addition, IFPRI research found that the ration shops were very ineffective in reaching poor people. It replaced them with a better-targeted Food-for-Education program that not only improved the nutrition of poor rural children, but contributed significantly to increased school enrollment. In carrying out its research activities, the CGIAR seeks to strengthen national agricultural research systems in the developing countries.

In addition to collaboration on research projects, the IARCs help strengthen research and management capacity. For example, IFPRI's project in Bangladesh provided training to over 200 local food policy analysts. The International Service for National Agricultural Research (ISNAR), based in the Netherlands, is a CGIAR center that focuses exclusively on capacity strengthening. It helped to establish the Kenya Agricultural Research Institute and a new regional Central American System for Agricultural Technology, among others. In collaboration with other CGIAR centers, ISNAR has also worked directly with farmers in Burkina Faso to help them set priorities for the national research institution.

In China, scientists from the International Rice Research Institute (IRRI) in the Philippines, the oldest of the CGIAR centers, have provided high-yielding rice lines and varieties to Chinese researchers, and also helped their Chinese counterparts to establish the National Rice Research Institute in 1989. Since the 1980s, over 3,400 Chinese agricultural scientists have received training at CGIAR centers. Globally, CGIAR centers have provided training to 50,000 researchers, educators, and extension agents from the developing world since 1971. At present, eminent developing country agricultural scientists chair key CGIAR systemwide committees and sit on boards of research centers.

Impacts of CGIAR Research

CGIAR research has borne considerable fruit:

- The adoption of new plant varieties, agricultural know-how, and technologies developed at CGIAR research centers were all crucial in the doubling of developing country grain harvests in a few decades;
- Between 80 and 100 percent of the rice land in major rice producing countries in Asia and Latin America is planted to high-yielding varieties developed at CGIAR centers or varieties that used CGIAR varieties as parents; in Latin America, unit costs and prices have declined by 50 percent over the past 30 years, saving consumers some $500 million, while increasing incomes of small farmers;
- 80 percent of the wheat varieties planted in developing countries came from the International Maize and Wheat Improvement Center (CIMMYT), a CGIAR center in Mexico, with the additional output due to high yield worth nearly $2 billion a year;

— Research by the system has helped Asia greatly increase fresh water fish production;
— Scientists at IRRI and another CGIAR center, the West Africa Rice Development Association (WARDA), crossed African and Asian rices with tissue culture techniques to develop a strain that is both hardy and has exceptionally broad leaves, thereby providing biological weed resistance; this not only improved rice harvests in West Africa, but have reduced the time women rice farmers have to spend have to spend weeding, allowing them to spend more time in the caring activities that are essential for good child nutrition; and
— CGIAR research helped West and Central African maize production to increase threefold between 1981-96, enough to feed an additional 40 million people each year, at a value of $1. 2 billion.

The report of the Third External Review of the CGIAR, issued in 1998, stated that "investment in the CGIAR has been the single most effective use of official development assistance (ODA), bar none. There can be no long-term agenda for eradicating poverty, ending hunger, and ensuring sustainable food security without the CGIAR."

Structure and Governance

The CGIAR system has no constitution or binding rules. It operates on the basis of consensus, and any government or organization that supports the mission, is willing to help shape the research agenda, and provides a certain minimum of financial support to the centers may join. The Chairperson is a World Bank Vice President, who, with the assistance of a small secretariat, provides intellectual and 7 managerial leadership, helps to harmonize policies and programs, and encourages linkages between the system and its varied partners. A panel of independent scientists, the Technical Advisory Committee, reviews each center's research program, makes recommendations to the Group on resource allocation, and makes recommendations on systemwide issues. A separate Finance Committee oversees the efficiency of financial management within the system.

The Group appoints an Oversight Committee to assure that due care and diligence are exercised in center and system operations. In addition, individual centers and the system as a whole undergo periodic reviews of their program and management by external experts. Each center is a separate corporate entity with its own governing board, management, and research staff. The Consultative Group meets twice a year to review policies, assess the impact of research, adopt a research agenda, and pledge funds in support of that agenda. Members make voluntary contributions to individual centers as they choose. In 1999, the CGIAR budget totaled $340 million.

Reducing Poverty

It may not be obvious how agricultural research can contribute to poverty eradication. For many people in the industrialized world, the term "agricultural research" is likely to conjure up a corporate scientist (or corporate-funded university scientist) working on how to make high-technology, capital-intensive agriculture in the industrialized countries more profitable. Such a caricature is grossly misleading, particularly for developing countries. First, almost all agricultural research in developing countries is carried out by poorly paid researchers working in small, publicly funded institutions. Second, around 70 percent of poor and food insecure people reside in rural areas and depend directly or indirectly on agriculture for their livelihoods. Whether in rural or urban areas, poor people spend as much as 50-70 percent of their incomes on food.

Low productivity in agriculture is a major cause of poverty, food insecurity, and poor nutrition in low-income developing countries, resulting in low incomes for farmers and farm workers, high costs per unit of food produced and therefore high consumer prices, little demand for goods and services produced by poor nonagricultural rural households, and urban unemployment and underemployment. Small farmers in developing countries face pre- and post-harvest crop losses due to pests and droughts. These result in low and fluctuating yields, incomes, and food availability.

Low soil fertility and lack of access to reasonably priced plant nutrients, along with acid, salinated, and water logged soils and other abiotic factors also contribute to low yields, production risks, and degradation of natural resources. Too often, poor farmers must clear forest or farm ever more marginal land in order to cultivate crops. Many rural poor people live in resource-poor areas. Inadequate infrastructure and poorly functioning markets, together with lack of access to land, credit, and technical assistance, add to the impediments.

Clearly, research that aims at achieving productivity gains among small farmers has considerable relevance to food security and poverty alleviation. Such research would seek to make crop varieties higher yielding through such traits as drought- and salttolerance and pest resistance, and to improve livestock in ways that benefit small-scale husbandry. It is essential that new production technologies for poor farmers be environmentally-friendly as well as yield-increasing. Continued low productivity in agriculture not only contributes to gaps between food production and demand in poor countries, but also prevents attainment of the broadbased income growth and lower unit costs in food production needed to help fill such gaps and improve food security.

While efforts to improve longer-term productivity on small-scale farms, with an emphasis on staple food crops, must be accelerated, research and policies are also needed to help farmers, communities, and governments better cope with risks resulting from poor market integration, poorly functioning markets, climatic fluctuations, and other factors. High potential areas will remain a key source of expanded food production in the future and, by minimizing the need to exploit new lands, they will help to reduce pressures on the natural resource base.

At the same time, more research must be directed to the development of appropriate technology for sustainable intensification of agriculture in resource-poor areas, where environmental risks are severe. This research must focus on such problems as reducing soil erosion, capturing and utilizing more moisture in the soils, generating and recycling organic sources of plant nutrients, developing more diverse cropping systems, and better integrating livestock and trees into cropping systems.

The private sector is unlikely to undertake much of the research needed by small farmers in developing countries because it cannot expect sufficient economic gains to cover costs. Benefits to society from such research can be extremely large but will not be obtained without public investments. Most studies of the private and social rates of return to investment in agricultural research find that they exceed 20 percent per year, compared to long-run real interest rates of 3-5 percent for government borrowing. The data support the view that such investments are a particularly productive use of scarce public resources. Without government involvement, too little agricultural research and development would take place.

There is a "market failure" in that the market does not provide the private sector with the incentives to support the agricultural research that would benefit society. For instance, the private sector has few incentives to conduct research on the problems and needs of small-scale farmers growing food crops for home consumption, because market development costs are high, small-scale farmers have little capacity to pay for improved technologies, and gains from basic research are difficult to capture when others can "free-ride" on the investments.

There are large areas of agricultural research that will not be carried out by the private sector because of the public-good nature of many of the products of research. Policies must expand

and guide research and technology development to solve problems of importance to poor people. Research should focus on crops relevant to small farmers and poor consumers in developing countries, such as bananas, cassava, potatoes, yams, sweet potatoes, grain legumes such as beans, rice, maize, wheat, and millet, along with livestock. International public goods research is critical, because, as in the above case of the cassava mealy bug, solutions from one part of the world may be widely applicable elsewhere.

More community participation is essential for setting research priorities. Unless these truly reflect and address the needs and problems of small-scale farmers who make up the bulk of the farming population in developing countries, agricultural research will not be fully utilized for achieving food security. There are several ways to expand such research. The first is to allocate additional public resources to agricultural research, carried out by public agencies, that promises large social benefits. In addition, the public sector can expand private-sector research on behalf of poor people by converting some of the social benefits to private gains, e. g. , by offering to buy exclusive rights to newly developed technology and make it available either for free or for a nominal charge to small farmers.

The private research agency would bear the risks, as it does when developing technology for the market. This arrangement is similar to that recently proposed by Harvard University economist Jeffrey Sachs for developing a malaria vaccine for use in Africa. Other arrangements include joint ventures between public and private agencies that share the costs and benefits of research and commercialize products and services, the establishment of research foundations, national competitive funds, and farmer financing through levies on agricultural production. Research and technology alone will not drive agricultural growth. The full and beneficial effects of agricultural research and technological change will materialize only if government policies are conducive to and supportive of poverty alleviation and sustainable management of natural resources.

For example, IFPRI research has found that in Tamil Nadu State in India, the adoption of high yielding grain varieties meant not only increased yields and cheaper, more abundant food for consumers, but income gains for small and larger-scale farmers alike, as well as for non-farm poor rural households. Increased rural incomes contributed to nutrition gains. Because the Tamil Nadu state government has pursued active poverty alleviation strategies, including extensive social safety net programs and investment in agriculture, rural development, nutrition, and education, and a fair measure of equity in access to resources such as land and credit, the benefits were widely shared. Where increased inequality followed the adoption of modern crop varieties, this was not because of factors inherent in the technology, but rather a result of policies that did not promote equitable access to resources and development of human capital.

And even in these areas, rural landless laborers usually found new job opportunities as a consequence of increased agricultural productivity, particularly where appropriate physical infrastructure and markets developed. On the other hand, successful adoption of these crop varieties depended on access to water, fertilizer, and pesticides. Thus, inequality between well endowed and resource poor areas increased. In their continuing effort to bring appropriate scientific tools to bear on the problems facing poor farmers in developing countries, many CGIAR centers have turned in a limited way to modern agricultural biotechnology. They have adopted this technology to the extent that it proves useful to their ongoing research programs rather than for its own sake.

Growth of Modern Biotechnology

Modern agricultural biotechnology—including, but not limited to, genetic engineering and the breeding of transgenic crops—

offers great potential as an instrument for achieving food security for all. As the technology is quite new, as compared to classical Mendelian plant breeding, there are also potential risks that need to be carefully assessed. There are many potential benefits for poor people in developing countries. Biotechnology may help achieve the productivity gains needed to feed a growing global population, introduce resistance to pests and diseases without costly purchased inputs, heighten crops' tolerance to adverse weather and soil conditions, improve the nutritional value of some foods, and enhance the durability of products during harvesting or shipping.

Bioengineered products may reduce reliance on pesticides, thereby reducing farmers' crop protection costs and benefiting both the environment and public health. Biotechnology research could aid the development of drought-tolerant maize and insectresistant cassava, to the benefit of small farmers and poor consumers. The development of cereal plants capable of capturing nitrogen from the air could contribute greatly to plant nutrition, helping small farmers who often cannot afford fertilizers. Biotechnology may offer cost-effective solutions to micronutrient malnutrition, which currently affects over 2 billion people, such as vitamin A- and iron-rich crops. By raising productivity in food production, agricultural biotechnology could help further reduce the need to cultivate new lands and help conserve biodiversity and protect fragile ecosystems.

The potential risks include negative environmental effects, such as gene escape to weeds, harm to beneficial species and natural resources, and resistance in pests; harm to human health due to the presence of anti-biotic resistance marker genes and the cauliflower mosaic virus in some transgenic crops, as well as the potential for allergen or carcinogen transfers; ethical concerns, such as whether humans have a right to cross species boundaries imposed in nature and the creation of foods that might violate religious dietary laws; and equity concerns.

Research and development in agricultural biotechnology, especially in the transgenic field, has been overwhelming dominated by a handful of large "life sciences" companies. They have focused almost exclusively on crops produced by industrialized country farmers or for industrialized country markets. The biggest risk of modern biotechnology for developing countries is that technological development will bypass poor people. A form of what Ismail Serageldin, the Chairperson of the CGIAR, calls "scientific apartheid" may well develop, in which cutting edge science becomes completely oriented toward industrial countries and large-scale farming. Modern biotechnology is not a silver bullet for achieving food security, but, its promise in the fight against poverty must be explored, and useful applications should be made available to poor farmers and consumers.

Utilizing this tool effectively will require the existence of an appropriate regulatory framework to assure food safety, biosafety, and antitrust enforcement. Financial and technical assistance to create such a framework in developing countries is urgently needed. Biotechnology has the potential to help enhance agricultural productivity in developing countries in a way that further reduces poverty, improves food security and nutrition, and promotes sustainable use of natural resources. Solutions to the problems facing small farmers in developing countries will benefit both farmers and consumers.

CGIAR BIOTECHNOLOGY ACTIVITIES

The CGIAR has not arrived at a systemwide consensus on biotechnology, and the use of molecular biology-based research methods at CGIAR centers is quite modest. The system's use of transgenic methods is even more limited. In 1998, the 16 centers endorsed a statement calling for "prudent application of the full range of biotechnology tools to achieve substantial and sustainable growth in agricultural productivity in poor countries. " The statement explicitly mentioned molecular markers, genetic engineering, and recombinant vaccines. It also stressed the potential environmental benefits of biotechnology, and identified the CGIAR centers as important intermediaries in ensuring access to technology and the development of effective biosafety regulations in developing countries.

A 1999 survey found that 13 of the 16 centers conducted some form of biotechnology research (including policy research at both IFPRI and ISNAR and the latter's program that helps developing countries develop biosafety systems). The largest program in both relative and absolute terms is that of the International Livestock Research Institute (ILRI) in Kenya. The $6. 5 million program accounts for over 23 percent of the center's budget, and supports work on characterization, conservation, and use of animal genetic resources; development of disease resistant livestock; and immunology and vaccine development.

Biotechnology is used in breeding programs at IRRI for rice; CIMMYT for both of its mandate crops; the International Crops Research Institute for the Semi-Arid Tropics (ICRISAT) in India for sorghum, millet, pigeon peas, chickpeas, and groundnuts; the International Center for Agricultural Research in Dry Areas in Syria for barley, wheat, lentils, and chickpeas; the International Potato Center (CIP) in Peru for potato and sweet potato; IITA for cassava, maize, plantains, bananas, yams, cowpeas, and soybeans;

WARDA for rice; and IPGRI for bananas and plantains. Budget shares range from a low of 1. 5 percent at WARDA to nearly 12 percent at IRRI. All of these centers continue to devote the bulk of their research resources to conventional breeding and other non-biotechnology work, and most give a high preference to conventional methods. The approach to biotechnology is pragmatic in that the centers utilize it where it offers time, cost, or reliability advantages as an integral part of ongoing germplasm improvement efforts. Those efforts - using both biotechnology and conventional techniques - focus on traits that will empower poor farm communities to build onto new technologies and have unbridled access to them. IRRI, for example, has used biotechnology to develop a salt-tolerant rice variety for coastal areas of the Philippines.

Centers also use biotechnology to achieve goals not possible by traditional methods. IRRI was able to pyramid multiple resistance genes to achieve more durable protection against the rice bacterial blight pathogen by using cost-effective DNA marker technology. The centers have not been involved in basic biotechnology research, focusing instead on applying basic work done elsewhere to the specific problems facing developing country farmers. CGIAR centers have played an important role in adapting existing technologies to developing country research conditions. An exception to the focus on applied research is CIMMYT's work on apomixis, i. e. , developing hybrid plants that are genetically capable of asexual reproduction. Successful results of this work would mean that farmers could reuse their own seed without the yield reduction associated with the reuse of second-generation seeds from hybrids. This would free farmers from having to purchase hybrid seeds every season, making high yield technology more accessible to poor farmers.

Apomixis is the antithesis of the so-called Terminator technology, which produces a sterile second generation as a means of protecting intellectual property rights. The CGIAR's total investment in biotechnology presently runs at under $25 million annually, or about 7. 7 percent of the total budget. This sum pales in comparison to the hundreds of millions of dollars that the life sciences companies spend on biotechnology research each year. Genetic engineering programs, in which genes are transferred from one species to another, are carried out at CIAT, CIMMYT, CIP, ICARDA, ICRISAT, IITA, ILRI, IPGRI, and IRRI. But the centers are engaged in many other forms of biotechnology research:

— Genetic diversity studies using molecular markers and other molecular characterization;

— Mapping to identify the gene or genes associated with a particular trait, again often using molecular markers;

— Gene discovery, i. e. , the isolation, characterization, and analysis of novel gene sequences, with an emphasis on naturally occurring sequences in a particular species or its close crossable relatives;

— Tissue culture, which involves regenerating entire organisms from single cells, both to preserve genetic resources and to produce disease-free planting materials for farmers;

— Pathogen detection using DNA-based and other biotechnology techniques; and

— Vaccine development at ILRI for small-scale livestock producers.

Centers also offer training to developing country scientists in these techniques. Some centers have offered training in designing and implementing biosafety regulations, laboratory safety procedures, and field testing procedures for transgenic crop varieties.

Intellectual Property Issues

The roles of the CGIAR centers in ex situ conservation of germplasm, conventional plant breeding, and especially biotechnology have, in recent years, put the system squarely upon the horns of a dilemma in the highly controversial arena of intellectual property rights. On the one hand, centers need access to products and processes for research that may be subject to patents or other forms of intellectual property protection in order to maintain their freedom to operate. At the same time, some centers are facing pressure to protect their own intellectual property, at least for defensive purposes. As the global agricultural research environment becomes increasingly more proprietary, the CGIAR centers must continue to maintain free access to the fruits of their research for poor farmers in developing countries.

As centers move into biotechnology research, they will have to use techniques, equipment, materials, and information subject to intellectual property rights protection by others. So far, centers have acquired access through commercial licenses, formal agreements providing limited use rights for specific research, and informal arrangements. For example, CIP has collaborated with life sciences firms on developing transgenic potatoes that are pest- and disease-resistant. The companies include PGS, a Belgian firm that has been acquired by Aventis, and AXIS Genetics of the UK. The agreement with PGS specifies that CIP may not utilize the products from the research in certain developed countries, but permits non-exclusive freedom to operate in most developing countries.

In some instances, owners of proprietary technology have permitted research by CGIAR centers but not distribution of the products that might result. As small, noncommercial institutions, the centers cannot negotiate arrangements on the basis of expected profits. In 1999, the centers began an intellectual property audit, aimed at developing an inventory of both the system's intellectual property and the third-party property in use at centers. The audit will help the system devise a strategy for securing the legal right to use others' proprietary technologies, as well as to evaluate the value of its own intellectual property and what to do about it. The audit follows rejection by the CGIAR in 1998 of the external review's recommendation that the system become a single corporate entity able to hold intellectual property rights and in a stronger position to negotiate deals to use the property of others.

This recommendation was controversial in part because the system's germplasm collection has already been placed in trust through the agreement with FAO to keep it in the public domain. Centers have taken action (usually through informal negotiations) to ensure that those who use samples comply with the terms of the material transfer agreements and refrain from seeking patents on the material received. The application for patents is country specific, and most existing patents related to agricultural research are held in the industrialized countries. Thus, to the extent that patents are not taken out in a particular developing country, that country's national agricultural research system (NARS) is free to use the research processes and traits in further research, adaptation, and release to farmers.

An IARC located in the country would have the same freedom to operate locally, although patents impinge on the production of international public goods. Farmers in the country are free to use commercialized improved seed even though it may be patented in other countries. The country, however, will not be able to export commodities produced by such seeds to countries where patents are in effect. Private sector corporations are likely to take out patents only in countries where they expect a sufficiently large commercial demand for the patented product. Similarly, since patenting is a costly affair, holders of patents on specific aspects of the research process may limit their patent applications to a few countries where they can obtain significant economic gains from providing access to the patented processes.

Thus, while all members of the World Trade Organization are required to develop acceptable intellectual property rights regimes, many of the poorest and smallest developing countries may not be affected greatly by the rapid increase in patenting of agricultural research processes and outputs simply because the patent holders will not take out patents in those countries. An implicit market segmentation is developing where countries and agricultural commodities of little interest to the private sector may be free to use novel research processes and traits patented elsewhere.

Agreements should be reached between NARS in developing countries and the major private sector patent holders in the area of agricultural research. Such agreements would recognize market segmentation and make available the output of modern biotechnology for further research and adaptation to benefit poor farmers and poor consumers in developing countries.

Appropriate technology would become public goods in the segmented markets while remaining private goods in profitable markets. The CGIAR could help facilitate such agreements while continuing to collaborate with the NARS.

Meanwhile, CGIAR centers are facing increasing pressure to act on intellectual property rights issues. Recently, IRRI organized a conference on plant variety protection and the likely prospect of rice patenting legislation in Asia. In March 2000, the CIMMYT Board announced that it would, on a case-by-case basis, consider intellectual property protection for germplasm improvements made with non-trust materials, as well as databases, publications, and other intellectual property. The policy covers patents, licensing agreements, and partnerships. The move is purely defensive, aimed at preventing private firms from patenting the non-trust germplasm. By invoking protections when appropriate, CIMMYT will be able to assure that the fruits of its research remain freely available to developing countries.

Other centers, including CIP and CIAT, have established similar policies, and these fall within the principles the CGIAR has adopted on intellectual property and genetic resources. Critics charge that the real goal of these CGIAR centers is to be able to further joint ventures with life sciences firms, which will want to profit from research results in the industrial world even if they are willing to give them away in developing countries. The CGIAR has taken a stand against the use of "any genetic system designed to prevent seed germination," such as the so-called Terminator, as a means of protecting intellectual property rights.

Some developing country governments and many NGOs, argue that as public institutions, the centers should keep their innovations in the public domain. Moreover, many accessions into center genebanks represent the results of on-farm breeding efforts, but the farmers (many of whom may no longer be living) and communities that undertook the work are unlikely to be parties to the patents. The provisions of the Convention on Biological Diversity concerning national sovereignty over genetic resources might also come into play (unlike in the case of in-trust materials). Developing country research institutions have often collaborated on center research projects as well. However, other developing country governments and private firms strongly support the need for defensive patenting.

A strong case can be made that CGIAR centers should take out defensive intellectual property protection, and that protection of the centers' own property will create bargaining chips for obtaining access to third-party intellectual property. The case is strengthened by an agricultural research environment that is not only increasingly proprietary, but extremely competitive. And yet, a strong counterargument can also be made. Establishing and defending intellectual property rights claims are time consuming and expensive. A journey down this road would take resources away from agricultural research to benefit poor people in developing countries. There are also fears that once centers get into the business of taking out patents, they will inevitably focus on research with a high commercial value. That would change the character of the CGIAR system altogether.

Collaboration has always been the hallmark of CGIAR research. NARS in developing countries have been and remain critical partners in the work of the centers. IARCs not only provide training and institution building on behalf of the NARS, but research projects are usually carried out through collaborative activities. NARS have played a pivotal role in adapting new agricultural science and technology developed at the IARCs to local conditions. The doubling of the release of new rice and maize varieties between 1966 and 1985 was due in large part to improved scientific capabilities in the developing countries themselves. The CGIAR has supported the growth of regional federations of agricultural research institutions throughout the developing world.

Collaborative partnerships extend beyond the public agricultural research agencies of the developing countries. For example, IFPRI carries out its research in more than 50 developing countries and collaborates with government ministries, agricultural research institutions, local and provincial governments, universities, local and international NGOs, donor agencies, regional intergovernmental organizations, and research institutions in the industrialized world. These partnerships can be multifaceted. The U. S. Agency for International Development provides funds to CGIAR centers for collaborative projects with U. S. universities. IFPRI has used these funds in some cases to help provide postgraduate training to researchers from developing countries.

Collaboration with Civil Society

Some 300 NGOs are engaged in collaborative research programs with CGIAR centers. Along with the NARS, NGOs play a critical role in ensuring that research results move from laboratories to farmers' fields. NGOs are also active in bringing the perspectives of small-farmers and other poor people to the table when research priorities are set. For example, in Maharashtra State in India, ICRISAT has worked with the National Academy of Agricultural Development and Entrepreneurship and Marketing, an Indian NGO that helps small farmers grow, market, and export their crops, to encourage farmers to adopt higher-yielding, quicker maturing varieties of groundnuts. IFPRI collaborated with Ghanian NGOs in studying of urban food security in Accra, with Honduran NGOs in looking at options for sustainable development of the hillsides, and with the Bangladesh Rural Advancement Committee on our microcredit research. CARE is a major collaborator on IFPRI's multicountry urban food security research.

CGIAR center partnerships with civil society also include direct work with farmers, engaging them in the design and implementation of research projects, so that they are not merely end users of the results. ICRISAT has worked with the Southern

African Development Community, a subregional intergovernmental body, on sorghum and millet improvement. Zimbabwean farmers have been involved in the initial selection of germplasm lines for breeding improved varieties. These partnerships are of considerable benefit to ICRISAT and NARS breeders, who can learn the desires and preferences of farmers as an integral part of the research process. Together with IITA and ILRI, ICRISAT has also collaborated with the Nigerian Institute of Agricultural Research on a farmer participatory research project aimed at integrating crop and livestock systems.

The design of the research project emerged from a meeting among farmers, extension agents, and scientists, and farmers managed experiments on appropriate croplivestock systems for their growing and living conditions. Obviously, the CGIAR's in situ conservation efforts can only succeed through partnerships with farmer and community-based organizations. CGIAR centers also actively engage NGOs in seminars, dialogs on policy issues, and consultations. In 1995, NGOs organized a "Global Biodiversity Forum" preceding the Second Conference of the Parties to the Convention on Biodiversity, held in Indonesia. The Indonesia-based Center for International Forestry Research, a CGIAR center, organized a session on forests and biodiversity. NGO officials participate in the Steering Committee of the Asian Rice Biotechnology Network, coordinated by IRRI.

The CGIAR faces critical choices in the near future. The freedom to operate issue will need to be addressed, particularly if the system is going to be able to conduct international public goods research to benefit poor farmers and consumers. Increasingly, not only products, but also gene sequences and research processes are subject to intellectual property rights protection. How to maintain a public interest character while remaining a player in a proprietary research environment will be a difficult balancing act.

A number of additional constraints will make the task even more difficult. A critical problem is underinvestment in agricultural research in developing countries, despite the need for productivity gains among poor farmers and the high social rates of return. Currently, low-income developing country governments grossly underinvest in agricultural research: less 0. 5 percent of the value of their agricultural production, compared to 2 percent in higher-income countries.

Developing countries have less than 100 agricultural researchers per one million economically active persons in agriculture, compared to 2,500 in industrialized counties. Sub-Saharan Africa, which desperately needs productivity increases in agriculture, has only 42 agricultural researchers per million people engaged in agriculture. Annual growth of African agricultural research expenditures declined in the 1970s, and the decline accelerated in the 1980s.

In Latin America, average public agricultural research budgets fell by 13 percent between the early 1980s and early 1990s. IFPRI research shows that even minor increases in aid to agricultural research for developing countries can significantly accelerate food supplies while relatively small cuts could have very serious negative effects. Donors of ODA have failed to adequately support agricultural research in the developing countries. According to FAO, total ODA declined about 10 percent in real terms between 1986-88 and 1995-97. Over this same period, ODA to agriculture dropped nearly 50 percent in real terms, and the share of aid to agriculture in total ODA fell from 25 percent to 14 percent, despite the rural center of gravity of poverty in developing countries. Aid to agriculture in the poorest continues to receive a lower priority from donors than aid to other (often complementary) sectors, such as child survival, family planning, and environmental protection, and aid to higher income developing countries and transition countries.

FAO has also found that aid commitments to agricultural research, training, and extension fell 52 percent in nominal terms over 1991-94. Support for the CGIAR alone has been flat since 1990. A possibly related problem is growing public opposition to the use of modern science to solve food and agriculture problems. This is especially true in the industrialized countries, particularly in Europe in relation to biotechnology, but the opposition is spreading to the developing world as well.

In the Philippines, environmentalists and some farm organizations staged protests during IRRI's 40th anniversary in April 2000. These groups joined the attack on both conventional and transgenic plant breeding, charging that IRRI had promoted "the U. S. agenda on counterinsurgency and corporate domination of domestic agricultural production," and that its "much-flaunted Green Revolution caused massive loss of biological diversity in rice paddies throughout Asia." The protesters opposed work on both hybrid rice and biotechnology by IRRI and the Philippine Rice Research Institute. Meanwhile, inside its campus, IRRI provided visitors with a display of over one hundred new rice varieties released to farmers around the world. Many of these varieties contain genes from over 20 different parental lines, using biodiversity to benefit poor people.

Globally, however, consumers' risk/benefit calculation concerning how their food is produced is likely to vary according to how they earn their income and how much of their income they spend on food. Consumers outnumber farmers by a factor of more than 20 in the European Union, and Europeans spend only a tiny fraction of their incomes on food. In the United States, farms account for less than 2 percent of all households, and the average consumer spends less than 12 percent of income on food. In the developed countries, consumers can afford to pay more for food, increase subsidies to agriculture, and give up opportunities for better-tasting and better-looking food. In contrast, in developing countries, poor consumers depend heavily on agriculture for their livelihoods and spend the bulk of their income on food. For them, reduced risks of losing a crop, productivity gains, and lower unit costs and prices are a matter of life and death.

Strong opposition to GM food in the European Union has resulted in severe restrictions on modern agricultural biotechnology, including a three-year moratorium on approval of commercial use of new genetically modified agricultural products. The opposition is driven in part by perceived lack of consumer benefits, uncertainty about possible negative health and environmental effects, widespread perception that a few large corporations will be the primary beneficiaries, and ethical concerns. For developing countries, the issue is not just whether to grow transgenic crops.

A large share of the food imported by many developing countries originates in the United States, where such crops are already widely planted. These importing countries must take a position on not only biosafety and food safety for their domestic production, but also whether they wish to insist on product differentiation and labeling in the case of imported food, which is likely to raise prices. Agricultural biotechnology research in developing countries could not expect to receive any scientific or financial support for their own research in this area. In practical terms, this would likely preclude most such research in the developing world, except for a few large countries such as China, India, and Brazil.

CHAPTER 13

FUTURE VISION OF AGRICULTURAL GROWTH

India has made impressive strides on the agricultural front during the last three decades. Much of the credit for this success should go to the several million small farming families that form the backbone of Indian agriculture and economy. Policy support, production strategies, public investment in infrastructure, research and extension for crop, livestock and fisheries have significantly helped to increase food production and its availability. During the last 30 years, India's foodgrain production nearly doubled from 102 million tons in the triennium ending 1973 to nearly 200 million tons (mt) in the triennium ending (TE) 1999. Virtually all of the increase in the production resulted from yield gains rather than expansion of cultivated area.

Availability of foodgrains per person increased from 452 gm/capita/day to over 476 gm/capita/day, even as the country's population almost doubled, swelling from 548 million to nearly 1000 million. Increased agricultural productivity and rapid industrial growth in the recent years have contributed to a significant reduction in poverty level, from 55 percent in 1973 to 26 percent in 1998. Despite the impressive growth and development, India is still home to the largest number of poor people of the world. With about 250 million below the poverty line, India accounts for about one-fifth of the world's poor. Child malnutrition extracts its highest toll in this country. About 25% children suffer from serious malnutrition. More than 50 percent of the pre-school children and pregnant women are anemic. The depth of hunger among the undernourished is also high.

India has high population pressure on land and other resources to meet its food and development needs. The natural resource base of land, water and bio-diversity is under severe pressure. The massive increase in population (despite the slowing down of the rate of growth) and substantial income growth, demand an extra about 2.5 mt of foodgrains annually, besides significant increases needed in the supply of livestock, fish and horticultural products. Under the assumption of 3.5% growth in per capita GDP (low income growth scenario), demand for foodgrains (including feed, seed, wastage and export) is projected in the year 2020 at the level of 256 mt comprising 112mt of rice, 82mt of wheat, 39mt of coarse grains and 22mt of pulses. The demand for sugar, fruits, vegetables, and milk is estimated to grow to a level 33mt, 77mt, 136mt and 116mt respectively. The demand for meat is projected at 9mt, fish 11mt and eggs 77.5 billion (Table 1).

Table 1: Demand for Agricultural Commodities

Item	Achieved TE 1997-99			Demand in 2020 (million tons)		Yield target in 2020	
	Area (Million tons)	Production (Million ha)	Yield (Kg/ha)	LIG	HIG	LIG	HIG
Rice	42.2	85.7	1903	112.4	111.9	2664	2652
Wheat	26.2	69.1	2582	82.3	79.9	3137	3045
Coarse cereal	30.7	30.4	1041	38.9	37.3	1268	1214
Cereal	99.1	185.2	1814	233.6	229.0	2357	2311
Pulses	21.7	13.8	608	22.3	23.8	1029	1095
Food-grains	120.8	199.0	1595	255.9	252.8	2119	2092

Edible oil	28.6	6.4	269	10.8	11.4	379.7	399
Potato	1.2	21.6	17188	27.8	30.6	22279	24566
Vegetables	5.3	74.5	14204	135.6	168.0	25673	31812
Fruits	3.2	43.0	13437	77.0	93.6	24064	29259
Sugarcane Gur	3.7	26.9	7006	32.6	33.7	8788	9088
Milk	-	71.2	-	115.8	137.3	-	-
Meat	-	5.0	-	8.8	11.4	-	-
Eggs number	-	2873	-	7750	10000	-	-
Fish	-	5.3	-	10.1	12.8	-	-

LIG: Low income growth 3.5% per capita GDP growth
HIG: High income growth 5.5% per capita GDP growth
Demand includes export 4.7mt rice, 3.6 mt wheat, and vegetables 2.2 mt fruits 1.4mt And fish 0.49 mt.

Future increases in the production of cereals and non-cereal agricultural commodities will have to be essentially achieved through increases in productivity, as the possibilities of expansion of area and livestock population are minimal. To meet the projected demand in the year 2020, country must attain a per hectare yield of 2.7 tons for rice, 3.1 tons for wheat, 2.1 tons for maize, 1.3 tons for coarse cereals, 2.4 tons for cereal, 1.3 tons for pulses, 22.3 tons for potato, 25.7 for vegetables, and 24.1 tons for fruits. The production of livestock and poultry products must be improved 61% for milk, 76% for meat, 91% for fish, and 169% for eggs by the year 2020 over the base year TE 1999. Average yields of most crops in India are still rather low.

EMERGING NEW TECHNOLOGIES

The agriculture sector recorded satisfactory growth due to improved technology, irrigation, inputs and pricing policies.

Livestock, poultry, fisheries and horticulture are surging ahead in production growth in recent years and will have greater demand in the future. Industrial and service sectors have expanded faster than agriculture sector resulting in declining share of agriculture in national accounts. Despite the structural change, agriculture still remains a key sector, providing both employment and livelihood opportunities to more than 70 percent of the country's population who live in rural areas. The contribution of small farmers to the national and household food security has been steadily increasing. The water availability for agricultural uses has reached a critical level and deserves urgent attention of all concerned.

India has high population pressure on land and other resources to meet its food and development needs. The natural resource base of land, water and bio-diversity is under severe pressure. Food demand challenges ahead are formidable considering the non-availability of favourable factors of past growth, fast declining factor productivity in major cropping systems and rapidly shrinking resource base. Vast uncommon opportunities to harness agricultural potential still remain, which can be tapped to achieve future targets. There are serious gaps both in yield potential and technology transfer as the national average yields of most of the commodities are low, which if addressed properly could be harnessed.

Concentration was on enhanced production of a few commodities like rice and wheat, which could quickly contribute to increased total food and agricultural production. This resulted in considerable depletion of natural resources and the rainfed dry areas having maximum concentration of resource poor farmers remained ignored, aggravating problems of inequity and regional imbalances. This also led to a high concentration of malnourished people in these rainfed, low productive areas. This era also witnessed rapid loss of soil nutrients, agro-biodiversity including indigenous land races and breeds.

The agriculture policy must accelerate all-round development and economic viability of agriculture in comprehensive terms. Farmers must be provided the necessary support, encouragement and incentives. It must focus both on income and greater on-farm and off-farm job and livelihood opportunities.

Issues for Sustainable Agricultural Development

In national priority setting, the following recurring and emerging issues for sustainable agricultural development and poverty alleviation must be considered:

(i) Population pressure and demographic transition;

(ii) Resource base degradation and water scarcity;

(iii) Investment in agriculture, structural adjustment and impact on the poor;

(iv) Globalization and implication on the poor;

(v) Modern science and technology and support to research and technology development; and

(vi) Rapid urbanization and urbanization of poverty, and deceleration in rural poverty reduction.

In addressing the above issues, a policy statement on agriculture must take note of the following uncommon opportunities:

— Conservation of natural resources and protection of environment.

— Vast untapped potential of our soil and water resources, and farming systems

— Technology revolution especially in the areas of molecular biology, biotechnology, space technology, ecology and management.

— Revolution in informatics and communication and the opportunity of linking farmers, extension workers and scientists with the national and international databases

Vision

The Agriculture Policy document must articulate a clear vision on following few basic parameters of the agricultural sector around which a policy framework must be developed.

— *Organization of agriculture*: A clear long-term vision where inter-sectoral linkages are explicit.

— *Sustainability and natural resource management*: Prescription must lie in the domain of political economy. Otherwise, allocating funds for watershed development, agroforestry, soil conservation, and so on will not produce desired results.

— *Institutional change*: Policy document must spell out new approaches and new institutions free from the shackles of bureaucratic and self-help framework.

— *Investment priorities*: There is a need to develop a consensus on investment themes, priorities and policies. Policy document must lend strength to the claim for greater investment in rural areas, and also re-examine its programmes in the light of complementarities.

— *Incentives*: Document must articulate a clear vision on the incentive framework.

— Risk management

CHALLENGES, POLICIES AND STRATEGIES

Enhancing Yield of Major Commodities

Yield of major crops and livestock in the region is much lower than that in the rest of the world. Considering that the frontiers of expansion of cultivated area are almost closed in the region, the future increase in food production to meet the continuing high demand must come from increase in yield. There is a need to strengthen adaptive research and technology assessment, refinement and transfer capabilities of the country so that the existing wide technology transfer gaps are bridged. For this, an appropriate network of extension service needs to be created to stimulate and encourage both top-down and bottom-up flows of information between farmers, extension workers, and research scientists to promote the generation, adoption, and evaluation of location specific farm technologies. Ample scope exists for increasing genetic yield potential of a large number of vegetables, fruits as well as other food crops and livestock and fisheries products.

Besides maintenance breeding, greater effort should be made towards developing hybrid varieties as well as varieties suitable for export purposes. Agronomic and soil researches in the region need to be intensified to address location specific problems as factor productivity growth is decelerating in major production regimes. Research on coarse grains, pulses and oilseeds must achieve a production breakthrough. Hybrid rice, single cross hybrids of maize and pigeonpea hybrids offer new opportunities. Soybean, sunflower and oil palm will help in meeting future oil demands successfully. Forest cover must be preserved to keep off climatic disturbances and to provide enough of fuel and fodder. Milk, meat and draught capacity of our animals needs to be improved quickly through better management practices.

Integrated nutrient management: Attention should be given to balanced use of nutrients. Phosphorus deficiency is now the most widespread soil fertility problem in both irrigated and unirrigated areas. Correcting the distortion in relative prices of primary fertilizers could help correct the imbalances in the use of primary plant nutrients -nitrogen, phosphorus, and potash and use of bio-fertilizers. To improve efficiency of fertilizer use, what is really needed is enhanced location-specific research on efficient fertilizer practices (such as balanced use of nutrients, correct timing and placement of fertilizers, and, wherever necessary, use of micronutrient and soil amendments), improvement in soil testing services, development of improved fertilizer supply and distribution systems, and development of physical and institutional infrastructure.

Arresting deceleration in total factor productivity: Public investment in irrigation, infrastructure development (road, electricity), research and extension and efficient use of water and plant nutrients are the dominant sources of TFP growth. The sharp deceleration in total investment and more so in public sector investment in agriculture is the main cause for the deceleration. This has resulted in the slow-down in the growth of irrigated area and a sharp deceleration in the rate of growth of fertiliser consumption. The most serious effect of deceleration in total investment has been on agricultural research and extension. This trend must be reversed as the projected increase in food and non-food production must accrue essentially through increasing yield per hectare. Recognising that there are serious yield gaps and there are already proven paths for increasing productivity, it is very important for India to maintain a steady growth rate in total factor productivity.

As the TFP increases, the cost of production decreases and the prices also decrease and stabilise. Both producer and consumer share the benefits. The fall in food prices will benefit the urban and rural poor more than the upper income groups, because the former spend a much larger proportion of their income on cereals than the latter. All the efforts need to be concentrated on accelerating growth in TFP, whilst conserving natural resources and promoting ecological integrity of agricultural system. More than half of the required growth in yield to meet the target of demand must be met from research efforts by developing location specific and low input use technologies with the emphasis on the regions where the current yields are below the required national average yield.

Literacy had a positive and significant relation with crop productivity and a strong link exists between literacy and farm modernisation. A recent study, has shown that literacy emerged as an important source of growth in adoption of technology, use of modern inputs like machines, fertilisers, and yield. Recognising that in the liberalised economic environment, efficiency and growth orientation will attract maximum attention. Literacy will play a far more important role in the globalised world than it did in the past. Contribution of literacy, through TFP, will be substantial on yield growth and domestic supply. As future agriculture will increasingly be science-led and will require modern economic management, high return to investment on education is expected.

The investments that are good for agricultural growth-technology and its dissemination, rural infrastructure (roads), education and irrigation - amount to a 'winwin' strategy for reducing rural poverty by also increasing the non-farm economy and raising rural wages. Creating infrastructure in less developed areas, better management of infrastructure and introduction of new technologies can further enhance resource productivity and TFP. Generation and effective assessment and diffusion of packages of appropriate technologies involving system and programme based approach, participatory mechanisms, greater congruency between productivity and sustainability through integrated pest management and integrated soil-water-irrigationnutrient management, should be aggressively promoted to bridge the yield gaps in most field crops. Besides this, efforts must be in place to defend the gains and to make new gains particularly through the congruence of gene revolution, informatics revolution, management revolution and eco-technology.

Many observers have expressed concern that technological gains have not occurred in a number of crops, notably coarse cereals, pulses and in rainfed areas. Recent analysis on TFP growth based on cost of cultivation data does not prove this perception. In all the 18 major crops considered in the analysis, several states have recorded positive TFP growth. This is spread over major cereals, coarse grains, pulses, oilseeds, fibres, vegetables, etc. In most cases, in the major producing states, rainfed crops also, showed productivity gains.

There is thus strong evidence that technological change has generally pervaded the entire crop sector. There are, of course, crops and states where technological stagnation or decline is apparent and these are the priorities for present and future agricultural research. Farming system research to develop location specific technologies and strategy to make grey areas green by adopting three-pronged approach - watershed management, hybrid technology and small farm mechanisation will accelerate growth in TFP. It is necessary to enlarge the efforts for promoting available dry land technologies. Promoting efficient fertiliser practices, improving soil-testing services, strengthening distribution channel of critical inputs specially quality seed and development of physical and institutional infrastructure will help resource-poor farmers.

Bridging Yield Gaps: Vast untapped potential in the yield exists for all crops in most of the states accounting for more than three-fourths of crop area. Emphasis must be given to the states in which current yield levels are below the national average yield. Bihar, Orissa, Assam, West Bengal and Uttar Pradesh are the priority states accounting for 66% of rice area which need emphasis on bridging yield gaps to attain target demand and yield growth. For wheat we must focus mainly on Uttar Pradsh, Madhya Pradesh, Bihar and Rajasthan accounting for 68% of wheat area. For coarse cereals, major emphasis must be given to Rajasthan, Maharashtra, Karnataka, Madhya Pradesh, Andhra Pradesh and Uttar Pradesh.

To meet the demand for pulses greater emphasis is needed in almost all the states with particular focus on Madhya Pradesh, Maharashtra, Rajasthan, Gujarat, Andhra Pradesh, Karnataka and Uttar pradesh which have three-fourths of total pulse area. The target growth in pulse yield from these states annually must be 6 per cent; otherwise the nation will experience shortage of pulses for all times to come. The task of attaining self sufficient in pulses production looks difficult without area expansion and irrigation. In cases of oilseeds greater emphasis is needed on Andhra Pradesh, Madhya Pradesh, Rajasthan, Maharashtra, Karnataka, West Bengal and Uttar Pradesh to increase the yield by about 4 per cent. The possibilities of developing processing industry for extracting edible oils from non-oilseeds commodities, like rice bran etc, needing to be explored.

The introduction of palm cultivation for oil production may release pressure on traditional oilseeds crops to meet future edible oil demand. In case of sugarcane, research and development efforts are to be strengthened in Uttar Pradesh and Bihar to increase the yields per hectare by about 4% per annum. The demand for sugar can also be met by developing mini sugar mills so that substantial sugarcane production can be diverted from Khandsari to sugar production. This may also help release some sugarcane area to other crops. Cotton crop requires greater yield improvement emphasis on 81 per cent of the cotton area in Maharashtra, Gujarat and Andhra Pradesh.

Water for Sustainable Food Security

India will be required to produce more and more from less and less land and water resources. Alarming rates of ground water depletion and serious environmental and social problems of some of the major irrigation projects on one hand, and the multiple benefits of irrigation water in enhancing production and productivity, food security, poverty alleviation, as mentioned earlier, are well known to be further elaborated here: In India, water availability per capita was over 5000 cubic metres (m3) per annum in 1950. It now stands at around 2000 m3 and is projected to decline to 1500 m3 by 2025. Further, the quality of available water is deteriorating. Also, there are gross inequalities between basins and geographic regions.

Agriculture is the biggest user of water, accounting for about 80 percent of the water withdrawals. There are pressures for diverting water from agriculture to other sectors. A study. has warned that re-allocation of water out of agriculture can have a dramatic impact on global food markets. It is projected that availability of water for agricultural use in India may be reduced by 21 percent by 2020, resulting: in drop of yields of irrigated crops, especially rice, thus price rise and withdrawal of food from poor masses. Policy reforms are needed from now to avoid the negative developments in the years to come. These reforms may include the establishment of secure water rights to users, the decentralization and privatization of water management functions to appropriate levels, pricing reforms, markets in tradable property rights, and the introduction of appropriate water-saving technologies.

The needs of other sectors for water cannot be ignored. Therefore it is necessary that an integrated water use policy is formulated and judiciously implemented. Several international initiatives on this aspect have been taken in recent years. India should critically examine these initiatives and develop its country-specific system for judicious and integrated use and management of water. A national institution should be established to assess the various issues, regulatory concerns, water laws and legislations, research and technology development and dissemination, social mobilization and participatory and community involvement, including gender and equity concerns and economic aspects. This institution should function in a trusteeship mode and seen as the flagship of a national system for sustainable water security.

Rainfed Ecosystems

Resource-poor farmers in the rainfed ecosystems practice less-intensive agriculture, and since their incomes depend on local agriculture, they benefit little from increased food production in irrigated areas. To help them, efforts must be increased to disseminate available dry land technologies and to generate new ones. It will be necessary to enlarge the efforts for promoting available dry land technologies, increasing the stock of this knowledge, and removing pro-irrigation biases in public investment and expenditure, as well as credit flows, for technology-based agricultural growth. Watershed development for raising yields of rainfed crops and widening of seed revolution to cover oilseeds, pulses, fruits and vegetables. Farming system research to develop location specific technologies must be intensified in the rainfed areas. Strategy to make grey areas green will lead to second Green Revolution, which would demand three-pronged strategy - watershed management, hybrid technology and small farm mechanisation.

Diversification of Agriculture and Value Addition

In the face of shrinking natural resources and ever increasing demand for larger food and agricultural production arising due to high population and income growths, agricultural intensification is the main course of future growth of agriculture in the region. Research for product diversification should be yet another important area. Besides developing technologies for promoting intensification, the country must give greater attention to the development of technologies that will facilitate agricultural diversification particularly towards intensive production of fruits, vegetables, flowers and other high value crops that are expected to increase income growth and generate effective demand for food. The per capita availability of arable land is quite low and declining over time.

Diversification towards these high value and labour intensive commodities can provide adequate income and employment to the farmers dependent on small size of farms. Due importance should be given to quality and nutritional aspects. High attention should be given to develop post-harvest handling and agroprocessing and value addition technologies not only to reduce the heavy post-harvest losses and also improve quality through proper storage, packaging, handling and transport.

The role of biotechnology in post-harvest management and value addition deserves to be enhanced.

Post-Harvest Management, Value Addition and Cost-Effectiveness

Post-harvest losses generally range from 5 to 10 percent for non-perishables and about 30 percent for perishables. This loss could be and must be minimized. Let us remember, a grain saved is a grain produced. Emphasis should therefore be placed to develop post-harvest handling, agro-processing and value-addition technologies not only to prevent the high losses, but also to improve quality through proper storage, packaging, handling and transport. With the thrust on globalization and increasing competitiveness, this approach will improve the agricultural export contribution of India, which is proportionately extremely low.

Cost-effectiveness in production and postharvest handling through the application of latest technologies will be a necessity. The agro-processing facilities should preferably be located close to the points of production in rural areas, which will greatly promote off-farm employment. Such centres of processing and value addition will encourage production by masses against mass production in factories located in urban areas. Agricultural cooperatives and Gram Panchayats must play a leading role in this effort. In doing so, the needs of small farmers should be kept in mind.

Investment in Agriculture and Infrastructures

The public investment in agriculture has been declining and is one of the main reasons behind the declining productivity and low capital formation in the agriculture sector. With the burden on productivity - driven growth in the future, this worrisome trend must be reversed. Private investment in agriculture has also been slow and must be stimulated through appropriate policies. Considering that nearly 70 percent of India still lives in villages, agricultural growth will continue to be the engine of broad-based economic growth and development as well as of natural resource conservation, leave alone food security and poverty alleviation. Accelerated investment are needed to facilitate agricultural and rural development through:

— Productivity increasing varieties of crops, breeds of livestock, strains of microbes and efficient packages of technologies, particularly those for land and water management, for obviating biotic, a biotic, socio-economic and environmental constraints;
— Yield increasing and environmentally-friendly production and post-harvest and value-addition technologies;
— Reliable and timely availability of quality inputs at reasonable prices, institutional and credit supports, especially for small and resource-poor farmers, and support to land and water resources development;
— Effective and credible technology, procurement, assessment and transfer and extension system involving appropriate linkages and partnerships; again with an emphasis on reaching the small farmers;
— Improved institutional and credit support and increased rural employment opportunities, including those through creating agriculture-based rural agro-processing and agro-industries, improved rural infrastructures, including access to information, and effective markets, farm to market roads and related infrastructure;
— Particular attention to the needs and participation of women farmers; and
— Primary education, health care, clean drinking water, safe sanitation, adequate nutrition, particularly for children (including through mid-day meal at schools) and women.

These investments will need to be supported through appropriate policies that do not discriminate against agriculture and the rural poor. Given the increasing role of small farmers in food security and poverty alleviation, development efforts must be geared to meet the needs and potential of such farmers through their active participation in the growth process. Government should facilitate and support community level action by private voluntary organizations, including farmers groups aimed at improving food security, reducing poverty, and assuring sustainability in the management of natural resources.

In addition, governments should enhance efforts to ensure good nutrition and access to sufficient food for all through primary health care and education for all. Modern biotechnology tools, genetic engineering, as well as conventional breeding methods are all expected to play important roles in the generation of higher yielding, pest and stress resistant varieties of rice, wheat, maize and other cereal crops. The availability of genetic innovations in developing countries will depend on continued high levels of investments in agricultural research, both at the international and the national levels.

Free and unhindered access to germplasm to breeders worldwide is absolutely crucial to the rapid dissemination and adoption of improved germplasm. This free movement and the dissemination of modern biotechnology innovations to developing countries are hampered by increased patent protection and private sector investments. There is an urgent need to address this problem of free access to technology in the future. Increased attention will also have to be given to development of sustainable systems that protect the natural resource base. Recent evidence of resource degradation and declining productivity in some intensively cropped areas is of particular concern. Also population driven intensification of agriculture without the use of external inputs, is leading to a serious problem of mining soil fertility.

Sustaining global food supplies will depend on continued high levels of investments in research and technology development. It is essential that research capacity has to be increased substantially. In addition to investments in research, infrastructure investments, particularly in irrigation, transport and market infrastructure development are equally important for sustaining the productivity and profitability of food crop production. Mobilize the best of science and development efforts (including traditional knowledge and modern scientific approach) through partnerships involving national and international research institutions, NGOs, farmers' organizations and private sector in order to tackle the present and future problems of food security and production.

Fighting against Poverty and Hunger

Nearly one-fourth of India's population, 251 million out of nearly one billion, is below the poverty line. One hundred seventy millions of the poor, 68 percent, are rural and the remaining 32 percent are urban (Table 2). Number at the national level in rural area has decreased after 1983; the number of poor in the cities has been increasing. This is essentially due to migration of the destitute from villages to cities. There are serious implications of this trend on feeding the cities and food security of urban people, urban poverty and environment. A question may be asked as to whether the rural settings and opportunities could be improved for securing livelihood security and consequently rationalizing the migration to the cities.

Table 2. Number and percentage of population below poverty line in India

	Rural		Urban		All India	
Year	No. of persons (million)	% of person	No. of persons (million)	% of person	No. of persons (million)	% of person

1973	261	56	60	49	321	55
1983	254	46	74	41	328	45
1993	213	33	75	32	288	32
1998	170	24	81	30	251	26

An analysis of the incidence of rural poverty and hunger by farm size revealed that more than half of the landless people are poor. Poverty got significantly reduced from 54 percent in the landless group to 38 percent in the sub-marginal group (Table 3), suggesting that even a small piece of land, less than ½ hectare, can greatly reduce both poverty and hunger.

Table 3. Incidence of hunger and poverty by farm size in rural India

Land class	Percent of population	
	Hungry	Poor
Land less	49	54
<0.5	ha	32 38
0.5-1ha	24	27
1.0-2ha	17	19
2.0-4ha	12	14
>4ha	12	13

The incidence of hunger and poverty gets reduced as one is able to meet even part of his/her dietary energy requirement through growing his/her own food (Table 4). Studies show that even a small plot of one's own helps women to escape extreme poverty and deprivation. Land is the main asset for livelihood security. Although several factors affect the extent and depth of poverty and hunger, some of them have overwhelming impacts under the Indian setting. These include, irrigation, farming system and literacy.

Table 4: Relationship between home produced calories and hunger and poverty in rural India

Degree of home produced calories	Percent of population	
	Hungry	Poor
0	49	41
< 25 percent	36	34
25-50 percent	26	25
50-75 percent	23	20
>75 percent	17	29

Generally, there is higher concentration of poor, and hungry people in rainfed areas as compared with those in irrigated zones. Even with 20 percent of the irrigation intensity, there is a sharp fall in the proportion of hunger and poverty and it remains there irrespective of further intensification of irrigation (Table 5).

Table 5: Impact of irrigation on alleviation of hunger and poverty in India

Land class	Percent of population	
	Hungry	Poor
Rainfed	33	35
<20	20	22
20-50	22	23
50-80	18	24
>80	19	26

Evidences suggest that extensive irrigation will prove much more effective than to adding more and more water, and often wasting it along with the associated degradation of the natural resources. Such a policy will not only reduce poverty and hunger, but will also promote equity and environmental protection and natural resource conservation. An effective water policy and institutional support is needed to ensure judicious and equitable allocation, distribution and exploitation of water and water resources.

Livestock has the highest effect on reducing poverty and hunger. In rural India, 43 percent of the people who do not own even a single livestock are malnourished. Addition of one cattle or one buffalo to their assets reduces the hunger prevalence by 16 and 25 percentage points, respectively (Table 6). Only 14 percent of the people who owned one cattle and one buffalo were malnourished. In urban areas also, the addition of one cattle or one buffalo had significant impact on reduction of proportion of malnourished people. Livestock sector should also receive high priority with multiple objectives of diversifying agriculture, raising income and meeting the nutritional security of the poor farm households.

Table 6: Impact of livestock on alleviation of hunger and poverty in India

Livestock	Percent of population			
	Rural		Urban	
	Hunger	Poor	Hunger	Poor
None	36	28	43	55
Cow	31	25	29	42
Buffalo	26	18	20	33
Cow & buffalo	14	8	7	4

Literacy has a very high impact on poverty alleviation as well as on hunger reduction (Table 7). The illiterate people, whether urban or rural, are the most poor and malnourished. In urban areas the impact of literacy on poverty is the highest. Education, even above primary level, is extremely effective in reducing both poverty and hunger. Graduate and technical education is, of course, the most important instrument for reducing both poverty and hunger. But its impact is most visible on poverty reduction.

Table 7: Impact of literacy on alleviation of hunger and poverty in India

Literacy level	Percent of population			
	Malnourished		Below poverty level	
	Rural	Urban	Rural	Urban
Zero	36	28	43	55
Below Primary Level	31	25	29	42
Above Primary Level	26	18	20	33
Graduate and Technical	14	8	7	4

Therefore, the education policy of the country must be geared to remove illiteracy as soon as possible, as 50 percent of our people are still illiterate. Free education up to 8'th standard coupled with mid-day meal in the schools will go a long way in reducing both poverty and hunger and will thus help build a strong India. Further, this move will greatly reduce the violation of child labour laws and will offset some of the non-tariff restrictions imposed by developed countries on exports from developing countries on the grounds of use of child labour.

Empowering the Small Farmers

Contributions of small holders in securing food for growing population have increased considerably even though they are most insecure and vulnerable group in the society. The off-farm and non-farm employment opportunities can play an important role. Against expectation under the liberalized scenario, the non-agricultural employment in rural areas has not improved. Greater emphasis needs to be placed on non-farm employment and appropriate budgetary allocations and rural credit through banking systems should be in place to promote appropriate rural enterprises. Specific human resource and skill development programmes to train them will make them better decision-makers and highly productive. Human resource development for increasing productivity of these small holders should get high priority.

Thus, knowledge and skill development of rural people both in agriculture and non-agriculture sectors is essential for achieving economic and social goals. A careful balance will therefore need to be maintained between the agricultural and non-agricultural employment and farm and non-farm economy, as the two sectors are closely inter-connected. Raising agricultural productivity requires continuing investments in human resource development, agricultural research and development, improved information and extension, market, roads and related infrastructure development and efficient small-scale, farmer-controlled irrigation technologies, and custom hiring services. Such investments would give small farmers the options and flexibility to adjust and respond to market conditions.

For poor farm-households whose major endowment is its labour force, economic growth with equity will give increased entitlement by offering favourable markets for its products and more employment opportunities. Economic growth if not managed suitably, can lead to growing inequalities. Agrarian reforms to alleviate unequal access to land, compounded by unequal access to water, credit, knowledge and markets, have not only rectified income distribution but also resulted in sharp increases in productivity and hence need to be adopted widely. Further, targeted measures that not only address the immediate food and health care requirements of disadvantaged groups, but also provide them with developmental means, like access to inputs, infrastructure, services and most important, education should be taken.

Identification of need-based productive programs is very critical, which can be explored through characterisation of production environment. Improved agricultural technology, irrigation, livestock sector and literacy will be most important instruments for improving the nutritional security of the farm-households. Watershed development and water saving techniques will have far reaching implications in increasing agricultural production and raising calorie intake in the rainfed areas. Livestock sector should receive high priority with multiple objectives of diversifying agriculture, raising income and meeting the nutritional security of the poor farm households.

Need based and location-specific community programs, which promise to raise nutritional security, should be identified and effectively implemented. Expansion of micro credit programmes for income-generation activities, innovative approaches to promote family planning and providing primary health services to people and livestock and education should enhance labour productivity and adoption of new technologies. Development of the post-harvest sector, co-operatives, roads, education, and research and development should be an investment priority.

A congenial policy environment is needed to enable smaller holders to take the advantage of available techniques of production, which can generate more incomes and employment in villages. For this poor farmer needs the support of necessary services in the form of backward and forward linkages. Small-mechanised tools, which minimise drudgery and do not reduce employment, but only add value to the working hours are needed to enhance labour productivity. Special safety nets should be designed and implemented for them. Can agricultural co-operatives internalise and galvanize these marginal and excluded people? Off-farm employment provided through co-operatives will go a long way in pulling them out of the state where poverty breeds poverty. Therefore, investment in the empowerment of the small landholders will pay off handsomely.

Let us create rural centres of production and processing by masses through co-operatives or empowerment of Gram Panchayats to promote co-operatives. This will improve efficiency of input and output marketing and give higher income. There is need to disseminate widely post-harvest handling and agro-processing and value addition technologies not only to reduce the heavy post-harvest losses but also improve quality through proper storage, packaging, handling and transport. Panchayati Raj institutions and co-operatives can play significant role in all these directions. Giving them power over the administration, as contemplated under the 73rd and 74th Amendment of the Constitution has not been implemented seriously so far in any of the states.

Disaster Management

The frequency and intensity of disasters such as floods, droughts, cyclones and earthquakes have increased in the recent years. The devastating earthquake in Gujarat has brought untold miseries to the whole state and caused a national disaster. Special effort should be made to develop appropriate technologies for increasing preparedness to predict and to manage the disasters. Effective and reliable information and communication systems, contingency planning and national and international mobilization of technologies and resources are a must. Experiences of other countries in prevention and management of the disasters should be shared.

Impact of Globalisation

The globalization of agricultural trade will bring to the fore access to markets; new opportunities for employment and income generation; productivity gains and increased flow of investments into sustainable agriculture and rural development. I believe that if managed well, the liberalization of agricultural markets will be beneficial to developing countries in the long run, It will force the adoption of new technologies, shift production functions upwards and attract new capital into the deprived sector. However, this will only come to pass if we are mindful of the interests of billions of small and subsistence oriented farmers, fisher-folk and forest dwellers in the short and medium tern. So far the magic of globalization has not been felt in India. During the past one-decade of liberalization certain trends such as deceleration of the growth rate of agricultural GDP, declaration in yield growth rates, and low non-agricultural employment have emerged against expectations. As we globalize, however, it is imperative that we do not forget social aspirations for a more just, equitable and sustainable way of life. Trade agreements must be accompanied by operationally effective measures to ease the adjustment process for a small farmer in developing countries.

BIBLIOGRAPHY

Allen, T. F. H., "Neolithic urban primacy: the case against the invention of agriculture". *Journal of Theoretical Biology*. 66:169-180, 1977.

Armelagos, George J.; Goodman, Alan H., and Jacobs, Kenneth H, "The origins of agriculture: Population growth during a period of declining health". *Population and Environment*. 1991; 13(1):9-22.

Asch, David L. and Asch Sidell, Nancy B., "Archaeological plant remains: applications to stratigraphic analysis". Hastorf, Christine A. and Popper, Virginia S., editors. *Current paleoethnobotony: analytical methods and cultural interpretations of archaeological plant remains*. Chicago: University of Chicago Press; 1988; pp. 86-96.

Blumler, Mark A. and Byrne, Roger, "The ecological genetics of domestication and the origins of agriculture". *Current Anthropology*.; 32(1):23-54, 1991

Boserup, Esther, *The Conditions of Agricultural Growth*, Chicago: Aldine; 1965.

Buikstra, Jane E.; Konigsberg, and Bullington, Jill, Fertility and the development of agriculture in the prehistoric midwest. American Antiquity. 51(3):528-546,1986.

Conklin, Harold C., *An ethnoecological approach to shifting agriculture*, Transactions of the New York Academy of Sciences, 2nd Series. 17:133-142, 1954.

Dirks, R., "Social responses during severe food shortages and famine". *Current Anthropology*.; 21:21-44, 1980

Gladwin, Hugh and Murtaugh, Michael, "The attentive-preattentive distinction in agricultural decision making". Bartlett, P. J. *Agricultural Decision Making*. New York: Academic Press; 1980.

McCorriston, Joy and Hole, Frank, "The ecology of seasonal stress and the origins ofagriculture in the Near East". *American Anthropolcgist.* 1991; 93:46-69.

Wilson, Wendy, "The Fulani model of sustainable agriculture: Situating Fulbe nomadism in a systemic view of pastoralism and farming", *Nomadic Peoples*. 1995; 36/37:35-52.

INDEX

--